FLUID MECHANICS FOR TECHNICIANS

FLUID MECHANICS FOR TECHNICIANS

Thomas B. Hardison, P.E.

Mechanical Engineering Technology
Catawba Valley Technical Institute

RESTON PUBLISHING COMPANY, INC.
Reston, Virginia
A Prentice-Hall Company

Library of Congress Cataloging in Publication Data

Hardison, Thomas B.
  Fluid mechanics for technicians.

  Includes bibliographical references and index.
  1. Fluid mechanics.   I. Title.
TA357.H355   620.1'06   76-20585
ISBN 0-87909-297-1

© 1977 by
Reston Publishing Company, Inc.
A Prentice-Hall Company
Reston, Virginia 22090

All rights reserved. No part of this book
may be reproduced in any way, or by any
means, without permission in writing
from the publisher.

10   9   8   7   6   5   4   3   2

Printed in the United States of America.

To Nancy and Jane

CONTENTS

PREFACE   xiv

LIST OF SYMBOLS AND ABBREVIATIONS   xv

chapter one   **PROPERTIES OF COMPRESSIBLE AND INCOMPRESSIBLE FLUIDS**   1

    1-1   Introduction, 1
    1-2   Fluids, 2
    1-3   Definitions of Fluid Properties, 2
    1-4   Systems of Units, 4
    1-5   Mass, Specific Weight, and Density, 4
            Questions and Problems, 6

chapter two   **VISCOSITY AND PASCAL'S LAW**   7

    2-1   Introduction, 7
    2-2   Viscosity, 7

2-3 Measurement of Viscosity, 9
2-4 Conversion of Viscosity Units, 10
2-5 Viscosity Index, 12
2-6 Pascal's Law, 12
2-7 Pressure in Compressible and Incompressible Fluids, 15
Questions and Problems, 16

chapter three

PROPERTIES PECULIAR TO INCOMPRESSIBLE FLUIDS 17

3-1 Introduction, 17
3-2 Surface Tension, 17
3-3 Surface Tension and Capillarity, 19
3-4 Vapor Pressure, 20
3-5 Compressibility of Liquids, 21
Questions and Problems, 23

chapter four

COMPRESSIBLE FLUIDS 25

4-1 Introduction, 25
4-2 Pressure Measurement, 25
4-3 Temperature, 28
4-4 Kinetic Theory of Gases, 30
4-5 Boyle's Law, 30
4-6 Charles' Law, 31
4-7 Perfect-Gas Law, 32
4-8 Universal Gas Constant, 32
4-9 Units of the Perfect-Gas Law, 33
4-10 Discussion and Problems, 33
Questions and Problems, 37

chapter five

THERMODYNAMIC PROCESSES OF COMPRESSIBLE FLUIDS 39

5-1 Introduction, 39
5-2 Energy Units, 40
5-3 Isothermal Process, 41
5-4 Adiabatic Process, 42
5-5 Polytropic Process, 43
5-6 Review, 43
Questions and Problems, 45

CONTENTS ix

chapter six   **FLUID STATICS**   **47**

     6-1   Introduction, 47
     6-2   Pressure-Height Relationships, 47
     6-3   Pressure Head, 49
     6-4   Fluid Density and Pressure, 50
     6-5   Pressure-Height Relationship for Gas, 53
     Questions and Problems, 54

chapter seven   **PRESSURE MEASUREMENT**   **57**

     7-1   Introduction, 57
     7-2   Units of Measurement, 57
     7-3   Manometers, 59
     7-4   Bourdon-Tube Gage, 63
     7-5   Static Pressure Measurement Under Dynamic Conditions, 64
     Questions and Problems, 65

chapter eight   **FLUID DYNAMICS— INCOMPRESSIBLE FLUIDS**   **69**

     8-1   Introduction, 69
     8-2   Principle of Conservation of Mass, 70
     8-3   Principle of the Conservation of Energy, 72
     8-4   Energy Additions and Losses, 75
     8-5   Torricelli's Theorem, 78
     Questions and Problems, 78

chapter nine   **FLOW TYPES AND FRICTION LOSS—INCOMPRESSIBLE FLUIDS**   **81**

     9-1   Introduction, 81
     9-2   Laminar Flow, 82
     9-3   Turbulent Flow, 83
     9-4   Reynolds Number, 83
     9-5   Flow Type and Reynolds Number, 84
     9-6   Determination of Friction Loss, 85
     9-7   Use of Moody's Chart, 87
     Questions and Problems, 91

chapter ten — FLUID DYNAMICS—COMPRESSIBLE FLUIDS 93

- 10-1 Introduction, 93
- 10-2 Gas Volume at Standard Conditions, 94
- 10-3 Equation of Continuity—Compressible Fluids, 94
- 10-4 Bernoulli's Equation for Compressible Fluids, 95
- 10-5 Mach Number, 100
- *Questions and Problems,* 101

chapter eleven — FRICTION LOSS—AIR 103

- 11-1 Introduction, 103
- 11-2 Air-Flow Losses in Pipe—Harris Formula, 105
- 11-3 Air-Flow Losses in Tubing, 106
- *Questions and Problems,* 109

chapter twelve — FLUID ENERGY, WORK, AND POWER 111

- 12-1 Introduction, 111
- 12-2 Pump Horsepower, 112
- 12-3 Cylinder Horsepower, 115
- 12-4 Energy Input and Bernoulli's Equation, 116
- 12-5 Input, Output, and Efficiency, 117
- *Questions and Problems,* 119

chapter thirteen — IMPULSE AND MOMENTUM 121

- 13-1 Introduction, 121
- 13-2 Impulse Turbine, 123
- 13-3 Jet Impingement on Moving Surfaces, 127
- 13-4 Work Done on a Moving Surface, 128
- 13-5 Forces on Pipe Bend with Enlarging or Contracting Sections, 129
- *Questions and Problems,* 133

chapter fourteen — ORIFICES, VENTURIS, AND PITOT TUBES 135

- 14-1 Introduction, 135
- 14-2 Orifice Flow Characteristics, 136

CONTENTS   xi

14-3	Venturi Tubes, 138
14-4	Pitot Tubes, 139
	Questions and Problems, 142

chapter fifteen
FRICTION LOSSES IN FITTINGS, VALVES, AND OTHER DEVICES   145

15-1	Introduction, 145
15-2	Mathematical Methods, 145
15-3	Equivalent Length of Pipe, 147
15-4	K Factor, 149
	Questions and Problems, 151

chapter sixteen
FLUID POWER   153

16-1	Introduction, 153
16-2	Hydraulics and Pneumatics, 161
16-3	Special Applications, 161
	Questions and Problems, 163

chapter seventeen
SELECTING THE FLUID POWER SYSTEM   165

17-1	The Hydraulic System, 165
17-2	Pneumatic System, 167
17-3	Air-Oil System, 169
17-4	Fluid Considerations, 170
17-5	Conditioning of the Fluid, 171
	Questions and Problems, 172

chapter eighteen
FLUID POWER COMPONENTS— CYLINDERS   175

18-1	Introduction, 175
18-2	How the Cylinder Functions, 175
18-3	Cylinder Types, 177
18-4	Cylinder Mounting, 179
18-5	Cushioned Cylinders, 181
18-6	Unusual Cylinder Applications, 182
	Questions and Problems, 183

chapter nineteen   **FLUID POWER COMPONENTS—VALVES**   **185**

     19-1   Introduction, 185
     19-2   Directional Control Valves, 185
     19-3   Spool-Type Directional Control Valves, 187
     19-4   Application of Directional Control Valves, 190
     19-5   Check Valve, 192
     19-6   Pressure Control Valves, 192
     19-7   Flow Control Valves, 196
     19-8   Pilot Operation of Valves, 196
     Questions and Problems, 197

chapter twenty   **PIPING AND TUBING—INTENSIFIERS**   **201**

     20-1   Introduction, 201
     20-2   Pipe, 202
     20-3   Threads, 203
     20-4   Tubing, 203
     20-5   Hose, 204
     20-6   Fittings for Tubing and Hose, 205
     20-7   Intensifiers, 205
     Questions and Problems, 206

chapter twenty-one   **PUMPS**   **207**

     21-1   Introduction, 207
     21-2   Cavitation in Pumps, 208
     21-3   Gear Pumps, 209
     21-4   Vane Pumps, 210
     21-5   Piston Pumps, 211
     Questions and Problems, 213

chapter twenty-two   **MOTORS AND HYDROSTATIC TRANSMISSIONS**   **215**

     22-1   Introduction, 215
     22-2   Motor Torque and Power, 215
     22-3   Hydraulic Motors, 218
     22-4   Comparison of Hydraulic and Electric Motors, 220
     22-5   Air Motors, 220
     22-6   Hydrostatic Transmissions, 221
     Questions and Problems, 223

CONTENTS   xiii

chapter twenty-three   **ACCESSORY COMPONENTS**   **225**

     23-1   Introduction, 225
     23-2   Accumulator, 225
     23-3   Strainers and Filters, 227
     23-4   Specifying a Filter, 228
     23-5   Reservoirs, 230
              Questions and Problems, 232

chapter twenty-four   **GRAPHIC SYMBOLS FOR FLUID POWER DIAGRAMS**   **233**

     24-1   Introduction, 233
     24-2   Use of Lines, 236
     24-3   Symbols for Components, 237
     24-4   Other Symbols, 240
              Questions and Problems, 240

chapter twenty-five   **HYDRAULIC CIRCUITS**   **243**

     25-1   Introduction, 243
     25-2   Simple Pump Circuit, 243
     25-3   Variable-Output Pump Circuit, 244
     25-4   Motor-Driven Pump Circuit, 245
              Questions and Problems, 246

chapter twenty-six   **PNEUMATIC CIRCUITS**   **249**

     26-1   Introduction, 249
     26-2   Complete Pneumatic Circuit, 249
     26-3   Two-Cylinder Partial Circuit, 250
     26-4   Valve-Operated Partial Circuit, 251
     26-5   Four-Way Valve-Operated Partial Circuit, 252
              Questions and Problems, 252

appendix one   **SUPPLEMENTARY TABLES**   **255**

appendix two   **ANSWERS TO EVEN-NUMBERED PROBLEMS**   **263**

INDEX   **267**

PREFACE

This book has two primary purposes. One is to integrate the engineering theory of fluid mechanics with the down-to-earth practical world of fluid power. The second purpose is to make the theory and its applications understandable and palatable to the typical student in a two-year engineering technology curriculum. The book has, therefore, been written for technical institute and community college use.

Those portions of fluid mechanics theory which have applications primarily in fluid power have been used in developing the text. However, conventional fluid mechanics theory relating to static and flow systems of fluids has also been incorporated. Parts relating to the older civil engineering hydraulics concept and to aerodynamics have not been included.

Calculus is not used in the book. However, the student will benefit if he has had some calculus, and it is assumed that he is proficient with algebra and trigonometry. Courses in physics also will be helpful in the course.

SI metrics have been incorporated to an extent which appears compatible with present day usage. While SI metrics will undoubtedly be used more in the future, several manufacturers have indicated that their primary use of metrics is in the area of component and part dimensional characteristics.

The latter part of the book is devoted to the applications in fluid power. While most of these applications were furnished by fluid power manufacturers and other industry sources, some are from the author's own industrial experience of 18 years. Primary components are discussed and their functioning explained to the extent that space allows.

It is hoped that the student will obtain understanding and some degree of knowledge of the field by using this text.

THOMAS B. HARDISON

LIST OF SYMBOLS AND ABBREVIATIONS

This list of standard symbols and abbreviations has been included here for easy access by the student. The following symbols and abbreviations are used throughout the text.

SYMBOL OR ABBREVIATION	MEANING
a	Acceleration
A	Area
°C	Temperature in degrees Celsius
d	Distance or diameter
Delta (Δ)	Change in ....
Epsilon (ϵ)	Absolute roughness
f	Friction factor
F	Force or flow factor
°F	Temperature in degrees Fahrenheit
g	Acceleration of gravity
G	Weight flow rate (lb per sec)
h	Head or height
k	Constant of proportionality
K	Temperature in degrees kelvin
lb/in.² & psi	Pressure–pounds per sq in.
L	Length
m	Molecular weight
M	Mass
Mu (μ)	Absolute viscosity
n	Exponent for adiabatic and polytropic processes
N_R	Reynolds number
Omega (ω)	Angular velocity

SYMBOL OR ABBREVIATION	MEANING
psia	Lb per sq in. absolute-pressure
psig	lb per sq in. gage-pressure
P	Pressure
P_f	Pressure drop
Q	Volume rate of flow
R	Gas constant in ft lb/lb °R
°R	Temperature in degrees Rankine
Rho (ρ)	Density
S_g	Specific gravity
SSU	Saybolt Seconds Universal—viscosity
t	Time
Tau (τ)	Shear stress in viscosity
T	Temperature
Upsilon (v)	Kinematic viscosity
u	Relative velocity
U	Velocity—used in viscosity
v	Velocity
V	Volume
V_s	Specific volume
W	Weight
W_s	Specific weight
Y	Distance—used in viscosity
>	Symbol meaning "greater than"
<	Symbol meaning "less than"

chapter one

PROPERTIES OF COMPRESSIBLE AND INCOMPRESSIBLE FLUIDS

1-1 INTRODUCTION

Fluid mechanics can be described as a branch of applied mechanics that deals with the study of fluids, their properties, and their behavior. This text is primarily concerned with those aspects of fluid mechanics that lend themselves to use most frequently in industry. Some examples of applications of fluid mechanics are the shop compressed-air system, which exists in most plants; the hydraulic lift at the automobile service station; the vacuum chuck, which you will find on some machine tools; the hydraulically operated landing gear used on many aircraft; and the hydraulic shock absorber used on automobiles.

Industry has given the name *fluid power* to the field of practical application of fluid mechanics. Although it is the purpose of this text to introduce the reader to fluid power, it is equally important that you understand basic theory so that you may know how fluid mechanics is correctly applied to a specific job.

Physics deals with such terms as work, horsepower, force, velocity, energy, and friction. We shall also use these terms in fluid mechanics; in some cases they may be used in a somewhat different manner, but the basic meaning of the term remains the same.

1-2 FLUIDS

As differentiated from a solid, we can define a *fluid* as a substance (or a state of matter) that has mass but no definite shape. Since a fluid has no definite shape, it will take the shape of the vessel in which it is contained. We also have to go a step further and note the difference between liquid and gaseous fluids.

Fluids that are liquid under the conditions of use are said to be theoretically *incompressible*. Fluids that are gaseous under the conditions of use are *compressible*. Note that we restricted this definition so that the condition of the fluid is always considered; we do this because substances may change their state of matter under different temperature and pressure conditions. Water is a liquid at room temperature, a solid (ice) below freezing temperature, and above 212° F and at atmospheric pressure it is gaseous (steam). Another difference between liquids and gases is that, at a constant temperature, the volume of a liquid will not change, even though the vessel is changed; in a closed vessel, the gas volume will change so as to equal the volume of the vessel.

An ideal fluid is defined as one that has no resistance to flow. Since resistance to flow is also termed *viscosity*, the ideal fluid would have zero viscosity. In our study of fluid mechanics we will find that fluids do possess viscosity of sufficient magnitude that it cannot be neglected. Actually, most fluids behave similarly to the ideal fluids but not exactly like the ideal fluids. There are ways to compensate for this departure from the theoretical behavior, which we will go into later.

1-3 DEFINITIONS OF FLUID PROPERTIES

The important properties of fluids that apply to both liquids and gases are defined in this chapter. We shall cover separately those properties which pertain only to liquids and those which pertain only to gases.

Mass: "The property of a body which determines the effect of a force applied to it"[1]**; or "a quantity of matter."**[2]

We determine mass by measuring the weight of the object or the matter in question, but the mass is the weight divided by the acceleration

---

[1] David Halliday and Robert Resnick, *Physics* (New York: John Wiley & Sons, Inc., 1960), p. 70.

[2] *Handbook of Chemistry and Physics*, 49th ed. (Cleveland: The Chemical Rubber Co., 1968), p. 85.

Sect. 1-3 / DEFINITIONS OF FLUID PROPERTIES

of gravity:

$$M = \frac{W}{g} \tag{1-1}$$

where M = mass
W = weight
g = acceleration of gravity = 32.2 ft/s² in the English system of units and 9.81 m/s² in the SI (metric) system.

Specific weight: The weight per unit volume; for example, the specific weight of water for ordinary temperature variations is 62.4 lb/ft³.

Density: The mass per unit volume.

$$\rho = \frac{M}{V} \tag{1-2}$$

where ρ (rho) = density; may be referred to as mass density
M = mass
V = volume

Specific volume: Mathematically, the reciprocal of the specific weight.

$$V_s = \frac{1}{W_s} \tag{1-3}$$

where V_s = specific volume
W_s = specific weight

Specific gravity: The ratio of the weight or mass of any fluid to that of an equal volume of a substance taken as a standard; in the cases of liquids and solids, water is the standard at a temperature of 4° Celsius.[3]

In the case of gases, air free of carbon dioxide and hydrogen is frequently used as the standard; in practice, we rarely use the term to pertain to anything other than liquids or solids. In this case, then, mathematically the specific gravity is

$$S_g = \frac{W_{\text{substance}}}{W_{\text{water}}} \tag{1-4}$$

where S_g = specific gravity
$W_{\text{substance}}$ = specific weight of substance
W_{water} = specific weight of water

---

[3] The Celsius temperature scale was previously called centigrade.

4 PROPERTIES OF COMPRESSIBLE AND INCOMPRESSIBLE FLUIDS / Chap. 1

In the SI system specific gravity may be obtained using the density of water. At 4° Celsius this is 1000 kg/m^3. Thus, the specific gravity of a substance is equal to its density divided by 1000. For example, a fluid that has a density of 1320 kg/m^3 has a specific gravity of 1.32. Older metric systems, such as the CGS (centimeter-gram-second) system, defined density on a g/cm^3 basis. Water has a density of 1 g/cm^3 in this system.

1-4 SYSTEMS OF UNITS

Two systems of units are in use throughout the world. Most industrial countries have standardized on the *International System of Units* (SI), or *metric system*. In the United States the *English system of units* is used predominantly, but there is increasing use of the SI system. Consequently, we shall use both the English and the SI system in this text. A conversion table for changing from one system to the other will be found in Table II of the Appendix.

Some metric terms still in common use in fluid mechanics are from the older CGS system. Examples are the poise and the stoke, used in the description of viscosity. Because these terms are still widely accepted, they will be used here and their SI equivalent terms listed also.

1-5 MASS, SPECIFIC WEIGHT, AND DENSITY

Practically speaking, the *mass* of a substance is a property that we can determine by weighing it and then dividing the weight by the constant of acceleration of gravity. In the English system, the unit of mass is called the *slug* and is obtained as follows: Using W in pounds and g in ft/s^2,

$$M = \frac{\text{lb}}{\text{ft/s}^2} = \frac{\text{lb s}^2}{\text{ft}}$$

The units of the slug are thus: lb s^2/ft.

In the SI system, the standard unit of mass is the *kilogram*. Since kilogram is the primary definition of mass in the metric system, it does not reduce to units of force, length, and time as does the slug in the English system.

One of our definitions of mass stated that it was the property of a body that determines the effect of a force applied to it. In physics, one of the basic laws is that force is equal to mass multiplied by acceleration, or $F = Ma$. The mass of any fluid is also subject to this relationship, just as is any mechanical object.

The relation of specific weight and density to weight are described

Sect. 1-5 / MASS, SPECIFIC WEIGHT, AND DENSITY

well enough by their definitions so that further discussion should not be necessary. We should note that temperature will change the specific weight of fluids, in most cases only slightly. Consequently, the mass, density, specific volume, and specific gravity will change. This is caused by the expansion or contraction of the fluid that results from the temperature change. In all our calculations we shall neglect this slight change, unless otherwise stated, and use the values in Table I in the Appendix.

EXAMPLE

The reservoir of a hydraulic system (Fig. 1-1) has a 10-gal capacity and the estimated capacity of the remainder of the system is 1.2 gal. It is planned to use a hydraulic fluid of 0.91 specific gravity in the system. What is the total fluid weight?

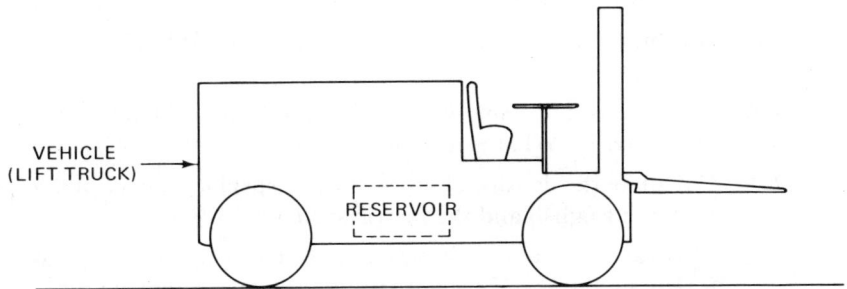

FIG. 1-1. Reservoir mounted in vehicle.

Solution

The total fluid volume $V = 10 + 1.2 = 11.2$ gal. Referring to Table II in the Appendix, $1\text{ ft}^3 = 7.48$ gal; then $V = 11.2 \text{ gal}/7.48 \text{ gal}/\text{ft}^3 = 1.5$ ft^3. If water is used, the weight W would be

$$W = 62.4 \text{ lb}/\text{ft}^3 \times 1.5 \text{ ft}^3 = 93.6 \text{ lb}$$

Since we are talking about the same volume of water and hydraulic fluid, we can say that the specific gravity S_g is

$$S_g = \frac{W_s}{W}; \quad \text{then } W_s = S_g W = (0.91)(93.6) = 85.2 \text{ lb}$$

W_s is, of course, the desired weight of the hydraulic fluid. The vehicle in which this hydraulic system is used is subject to an acceleration of 24 ft/s². What force would the fluid in the reservoir exert on its container due to this acceleration?

$$V_R = \text{volume of reservoir fluid} = \frac{10 \text{ gal}}{7.48 \text{ gal/ft}^3} = 1.33 \text{ ft}^3$$

$$W_R = \text{weight of the fluid} = (1.33 \text{ ft}^3)(62.4 \text{ lb/ft}^3)(0.91)$$
$$= 75.5 \text{ lb}$$

The resulting force is that from the relationship $F = Ma$ and is

$$F = \frac{75.5 \text{ lb}}{32.2 \text{ ft/s}^2}(24 \text{ ft/s}^2) = 56.3 \text{ lb}$$

QUESTIONS AND PROBLEMS

1-1.—A solid has a specific gravity of 2.2 and a volume of 2.3 ft^3. What is its weight?

1-2.—A volume of 3.2 ft^3 of a liquid weighs 164 lb. What is its specific volume?

1-3.—A fluid has a specific weight of 0.295 lb/in.^{3}. What is its specific gravity? What is its density?

1-4.—Calculate the density of an oil with a specific gravity of 0.97 in both the English and the metric systems.

1-5.—Mineral oil weighs 0.89 g/cm^3. What is its specific gravity? Calculate the specific weight in the English system of units.

1-6.—What force does a static body of fluid weighing 195 lb exert on the bottom of the container it is in? The container is cylindrical with axis vertical and is open to the atmosphere.

1-7.—Air at 60° F and atmospheric pressure has a density of 0.00237 slug/ft^3. Calculate its specific weight and specific volume.

1-8.—A fuel oil storage tank has a capacity of 250 gal. Calculate the weight of the fuel oil having a specific gravity of 0.892 required to fill the tank.

1-9.—A reservoir of a hydraulic system contains two immiscible liquids as shown in Fig. 1-2. Liquid A has a specific gravity of 1.595 and liquid B has a specific gravity of 0.90. Under static (no-flow) conditions and with the reservoir open to the atmosphere, which liquid will stay on top? With 5 gal of liquid A and 4 gal of liquid B, calculate the total weight.

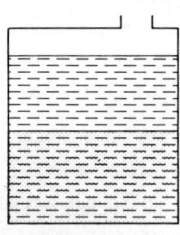

FIG. 1-2.

chapter two

VISCOSITY AND PASCAL'S LAW

$$\text{KINEMATIC } \nu = \frac{\text{ABSOLUTE}}{\text{DENSITY}} \; \frac{\mu}{\rho}$$

2-1 INTRODUCTION

Other properties common to both compressible and incompressible fluids are viscosity and a property that can be described by Pascal's law. *Pascal's law* states that a static fluid will transmit pressure equally in all directions, neglecting height.

2-2 VISCOSITY

Viscosity is not hard to visualize, but some of the mathematical terms used to describe it make it somewhat complicated. Many people think of the viscosity of a liquid as the weight of the liquid. This description of viscosity is not correct. A better definition of viscosity is as follows:

Viscosity: That property of a fluid which is a measure of its resistance to a shearing force.

Viscosity may also be thought of as a measure of the fluid's resistance to flow or of its internal friction. Figure 2-1 illustrates this resistance to a shearing force. Here there are two parallel plates with dimensions and

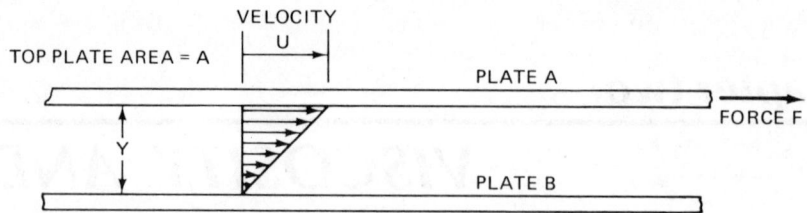

FIG. 2-1. Velocity gradient of fluid between two flat plates.

areas large enough so that the fluid which separates the plates is not affected by the edges of the plates. The top plate, A, is moving with a velocity U as shown, and the bottom plate, B, is stationary. The distance Y between the plates is small enough so that the plates can be considered as acting like a journal bearing with a film of lubricant between them. The force F moves plate A against the shearing resistance of the fluid. When this happens, particles of fluid at the surface of each plate will adhere to the plate and thus have the same velocity as the plate itself. If Y and U are not too great, the velocity gradient of the layers of fluid particles will be a straight line, as shown in Fig. 2-1. This means that, at a distance one half that of Y from plate B, the velocity will be one half that of U, and at other distances the velocity will vary proportionally.

In order to move in this manner, the layers of fluid particles have to slip with respect to each other, and this slippage requires a *shear stress*, τ (tau). From experiments it has been determined that the force F is proportional to AU/Y. If we assume a *constant of proportionality* μ (mu), we can say that $F = \mu AU/Y$. Also, since shear stress is equal to force/area, $\tau = F/A$ and $F = A\tau$. Then

$$F = \frac{\mu AU}{Y} = A\tau \quad \text{and} \quad \frac{\mu U}{Y} = \tau$$

Solving for μ,

$$\mu = \frac{\tau Y}{U}$$

The constant of proportionality μ is called the *absolute viscosity*. The units for absolute viscosity are:

$$\frac{\frac{\text{force}}{\text{area}}(\text{length})}{\text{length/time}} = \frac{\text{force/length}}{\text{length/time}} = \frac{\text{force time}}{\text{length}^2}$$

The most commonly used unit of viscosity is in the CGS system and is called the *poise*. One poise equals 1 dyn s/cm^2 (the *dyne* is a unit of force).

Because of the low values encountered for most fluids, the centipose (equal to 0.01 poise) is also used frequently. Conversion of the poise to the SI system gives an equivalent term called the *pascal second* (Pa·s), and 1 poise is equal to 0.1 Pa·s. The conversion can be made with the following SI unit equivalents:

1 pascal = 1 newton per square meter (N/m^2) (the *newton* is the basic unit of force in the SI system)

Substituting these terms for pascal,

$$1 \text{ Pa} \cdot \text{s} = 1 \frac{(N)(s)}{m^2}$$

This gives the basic units of force time/length^2 required for absolute viscosity. The complete conversion (1 poise = 0.1 Pa·s) is obtained when appropriate conversions from the dyne and centimeter are used.

Still another viscosity unit is *kinematic viscosity*. Kinematic viscosity is absolute viscosity divided by the fluid density ρ. In the CGS system the unit is called a *stoke* and is equal to 1 cm^2/s. The centistoke is also used and is equal to 0.01 stoke. The Greek letter upsilon, v, is used here to represent kinematic viscosity. The equivalent SI unit for kinematic viscosity is the meter squared per second, and 1 stoke = 10^{-4} m^2/s.

To review, there are two important viscosity terms, absolute viscosity and kinematic viscosity. The units for these are the poise for absolute viscosity and the stoke for kinematic viscosity in the CGS system. In the SI system the units are the pascal second and the meter squared per second.

Temperature affects the viscosity of fluids. In the case of liquids, an increase in temperature decreases the viscosity. The opposite is true for gases; an increase in temperature increases the viscosity of the gas. The viscosity of a gas is a property not used much in practice. However, the viscosity of a liquid and the change in viscosity due to temperature are important and often-used properties. Table III in the Appendix shows the viscosity values of different liquids at various temperatures.

2-3   MEASUREMENT OF VISCOSITY

The measurement of the viscosity of a liquid has to be done in the laboratory using special test equipment. One common piece of test equipment used for this purpose is the *Saybolt universal viscosimeter*. This instrument measures the time required for a certain volume of fluid to flow through a standard-size tube under a certain pressure head. The results of this test are in units of time (seconds) and are commonly referred to as *Saybolt seconds universal* (SSU).

Saybolt-universal-viscosimeter test results can be converted to kinematic viscosity, v, by the following equations: When the time t in seconds is between 32 and 100 seconds,

$$v = 0.00226t - \frac{1.95}{t} \qquad (2\text{-}1)$$

When $t > v$ (is greater than) 100 seconds,

$$v = 0.00220t - \frac{1.35}{t} \qquad (2\text{-}2)$$

In both these equations, v is in stokes.

Because of the change in viscosity with temperature, it is necessary to specify or note the temperature at which the testing is done. The commonly used temperatures for the Saybolt test are 70°, 100°, 130°, and 210° F. Thus, an SSU viscosity number should have a reference temperature noted with it.

We now have three viscosity terms with the names, symbols, units, and relationships shown in Table 2-1.

Table 2-1

Name	Symbol	CGS Unit	SI Unit	Relationship
Absolute viscosity	μ	Poise	Pascal second (Pa·s) (1 poise = 0.1 Pa·s)	Basic viscosity unit
Kinematic viscosity	v	Stoke	Meter squared per second (1 stoke = 10^{-4} m²/s)	Equals μ/ρ
Saybolt	SSU	Second		$v = 0.00226t - \frac{1.95}{t}$ $v = 0.00220t - \frac{1.35}{t}$

2-4 CONVERSION OF VISCOSITY UNITS

There is frequently a need to convert viscosities from the English system to the metric system and back again. To do this, the following relationships are helpful:

Sect. 2-4 / CONVERSION OF VISCOSITY UNITS

$$\text{For absolute viscosity:} \quad 1\,\frac{\text{lb s}}{\text{ft}^2} = 478.7 \text{ poises} \quad (2\text{-}3)$$

$$\text{For kinematic viscosity:} \quad 1\,\frac{\text{ft}^2}{\text{s}} = 929.03 \text{ stokes} \quad (2\text{-}4)$$

EXAMPLE 1

At 68°F the absolute viscosity of water is 1 centipoise. What is its corresponding value in English units?

Solution

From Eq. (2-3), 1 lb s/ft² = 478.7 poises; then

$$1 \text{ poise} = \frac{1 \text{ lb s}}{478.7 \text{ ft}^2} \quad \text{and} \quad 1 \text{ poise} = 2.08 \times 10^{-3}\,\frac{\text{lb s}}{\text{ft}^2}$$

Since 1 centipoise = $\frac{1}{100}$ poises, then

$$1 \text{ centipoise} = \frac{2.08 \times 10^{-3} \text{ lb s}}{100 \text{ ft}^2} = 2.08 \times 10^{-5}\,\frac{\text{lb s}}{\text{ft}^2}$$

EXAMPLE 2

Alcohol has a specific weight of 50 lb/ft³. At a given temperature its absolute viscosity is 2.49 × 10⁻⁵ lb sec/ft². Find its kinematic viscosity at the same temperature.

Solution

From Chapter One, density equals mass per unit volume. Since mass in the English system equals weight divided by gravity, the density ρ of alcohol is the specific weight divided by gravity:

$$\rho = \frac{50 \text{ lb/ft}^3}{32.2 \text{ ft/s}^2} = 1.552\,\frac{\text{lb s}^2}{\text{ft}^4}$$

$$v = \frac{\mu}{\rho} = \frac{2.49 \times 10^{-5} \text{ lb s/ft}^2}{1.552 \text{ lb s}^2/\text{ft}^4}$$

$$= 1.604 \times 10^{-5}\,\frac{\text{ft}^2}{\text{s}}$$

EXAMPLE 3

Convert the values of absolute and kinematic viscosities in Example 2 to CGS and SI system values.

Solution

From Eq. (2-3), 1 lb s/ft² = 478.7 poises and

$$\mu = 2.49 \times 10^{-5} \frac{\text{lb s}}{\text{ft}^2} \left(478.7 \frac{\text{poises}}{\text{lb s/ft}^2}\right)$$

$$= 1.192 \times 10^{-2} \text{ poise}$$

The kinematic viscosity $v = 1.604 \times 10^{-5}$ ft^2/s. From Eq. (2-4), 1 ft^2/s = 929.03 stokes. Then

$$v = 1.604 \times 10^{-5} \frac{\text{ft}^2}{\text{s}} \left(929.03 \frac{\text{stokes}}{\text{ft}^2/\text{s}}\right)$$

$$= 1.49 \times 10^{-2} \text{ stoke}$$

Table 2-1 is used to obtain the comparable SI conversions: 1 poise = 0.1 Pa·s and

$$1.192 \times 10^{-2} \text{ poise} = 1.192 \times 10^{-2}(0.1) \text{ Pa·s}$$

$$= 1.19 \times 10^{-3} \text{ Pa·s}$$

1 stoke = 10^{-4} m^2/s, and

$$1.49 \times 10^{-2} \text{ stoke} = (1.49 \times 10^{-2})(10^{-4}) \text{ m}^2/\text{s}$$

$$= 1.49 \times 10^{-6} \text{ m}^2/\text{s}$$

2-5 VISCOSITY INDEX

The *viscosity index* is an arbitrary method of stating the rate of change of viscosity of an oil with a change in temperature. The method is described in ASTM (American Society for Testing and Materials) standards and was set up to provide a means to rate different oils and their change in viscosity with temperature change. A high viscosity index of 100 means that the oil changes less with temperature. A low viscosity index of 20 means that the oil changes a greater amount with temperature.

The test was developed for lubrication oils and is generally used only for oils and hydraulic fluids. The viscosity index therefore is a restricted term applicable only to liquids.

2-6 PASCAL'S LAW

A fluid will take the shape of the vessel in which it is enclosed. If the fluid is a liquid, its volume will remain the same in the vessel, but if it is a gas in a closed container, its volume will change to the volume of the closed container. This is illustrated by Fig. 2-2. In Fig. 2-2(a) a clear container is filled with a liquid. It conforms to the container shape but keeps its original

Sect. 2-6 / PASCAL'S LAW

FIG. 2-2. Volumes of a liquid and a gas in a closed container.

volume. In Fig. 2-2(b) we can use smoke to illustrate that a gas occupies the entire volume of the container.

In Fig. 2-3 is a cylinder with two opposing pistons. The right-hand piston is spring-loaded and both pistons are assumed to be friction-free when moved in either direction horizontally. A fluid is contained in the cylinder. If we apply a force F to the left piston as shown, the reaction to force is provided by the fluid (no leakage of fluid is assumed).

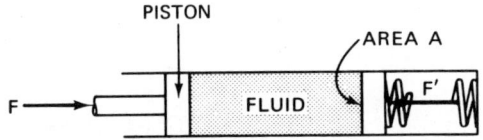

FIG. 2-3. Fluid under pressure.

By the laws of mechanics, there must be an equal and opposite force to F if the system is in equilibrium. The area that this equal and opposite force F' acts against is the area A of the spring-loaded piston. If the force F' is divided by the area A, the result is called the pressure P. Mathematically, the pressure $P = F'/A$. Stated in another manner, pressure equals force per unit area. Referring again to Fig. 2-3, this pressure is applied to the fluid itself by F', the reaction to the force F. It can then be said that the fluid is under a pressure P and that numerically $P = F/A$, since $F = F'$. Under static conditions the viscosity of the fluid will provide no frictional resistance to shear forces, as the viscosity only becomes a factor when layers of fluid are in relative motion.

Since the only other restraining force which acts on the fluid is that which is supplied by the walls of the vessel, it follows that the pressure P acts against the walls as well as the bottom and the piston. Further, in any given horizontal plane the pressure acting on any surface, whether horizontal, vertical, or inclined, is equal at all points. This is known as Pascal's law. Restated it is:

Pascal's law: Pressure at any one point in a static fluid is the same in all directions. NEGLECTING HEIGHT

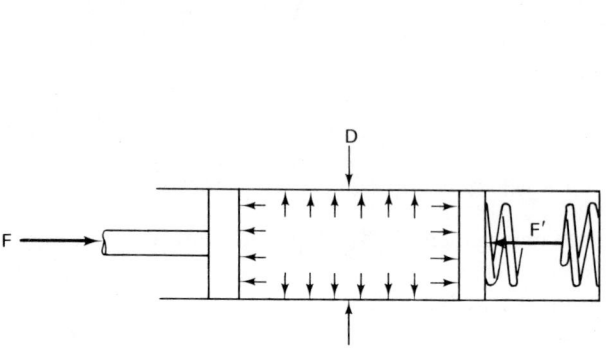

FIG. 2-4. Internal pressure distribution in a cylinder.

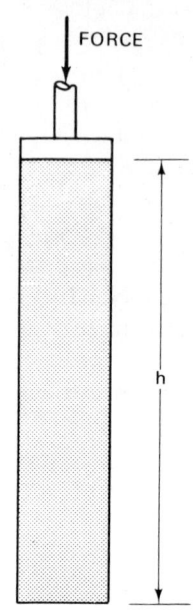

FIG. 2-5. Vessel with fluid under pressure and with elevation h of significant magnitude.

The cylinder of Fig. 2-3 is shown again in Fig. 2-4; here the internal pressure distribution is indicated by equal pressures at all points. We assume that diameter D is sufficiently small so that Pascal's law applies. If D is not small, the pressure varies as the elevation varies. In Fig. 2-5 is shown a vertical cylinder whose elevation is represented by height h. The pressure

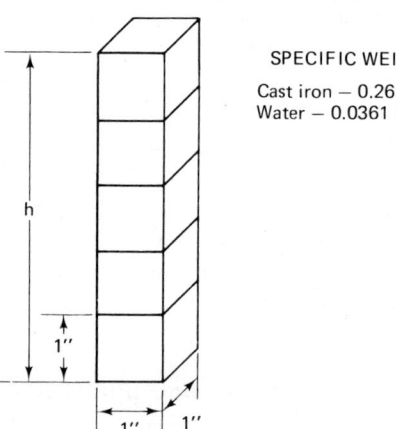

SPECIFIC WEIGHTS
Cast iron — 0.261 lb/in.^{3}
Water — 0.0361 lb/in.^{3}

FIG. 2-6. Unit pressure increase caused by elevation.

at the bottom of the cylinder is greater than the pressure at the piston. The difference in pressure is proportional to h.

This can be visualized by reference to Fig. 2-6. Imagine five cubes, each consisting of a 1-in.³ volume of cast iron stacked as shown. One cubic inch of cast iron weighs 0.261 lb. With only one cube, the pressure on the bottom surface is 0.261 lb/in.² (pressure may also written as psi, another abbreviation for pounds per square inch). With five cubes the pressure is $5 \times 0.261 = 1.305$ psi. This can be said since the area of the bottom is 1 in.². Therefore, if $h =$ height, the pressure is proportional to h.

Now imagine that, instead of the iron, there is an equal volume of a fluid such as water and that this is contained by a vessel. The pressure at the bottom is equal to five times the weight of 1 in.³ of water. Since 1 in.³ of water weighs 0.0361 lb, the pressure is $5 \times 0.0361 = 0.1805$ psi.

2-7 PRESSURE IN COMPRESSIBLE AND INCOMPRESSIBLE FLUIDS

There is no fundamental difference in measuring or using pressure in a compressible or an incompressible fluid system. A practical difference may exist, however, if the value of h, the height, becomes high enough to result in significant pressures in an incompressible (liquid) system. In Fig. 2-7(a) the cylinder is filled with a gas (compressible fluid) and in part (b) with a liquid (incompressible fluid). When the force F is first applied in part (a), the height h changes appreciably as the piston moves down and compresses the gas. It stops at the final position h', when the reaction from the internal pressure equals F. In part (b) h remains the same because the liquid is incompressible.

Because the weight of the gas in most cases is very small compared to the weight of any liquid, it can usually be neglected. For example, air

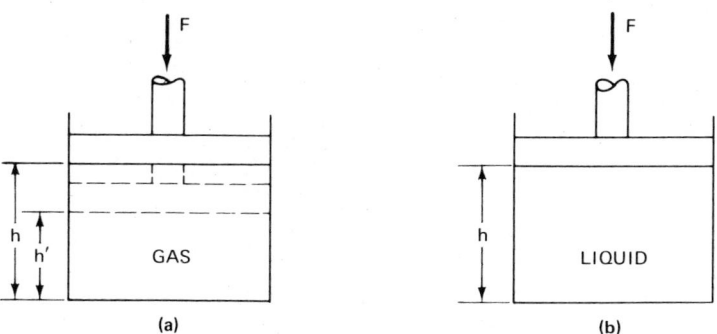

FIG. 2-7. Effect of compressibility on piston travel.

at atmospheric pressure and 68° F weighs 0.0752 lb/ft^3, compared to water, which weighs 62.4 lb/ft^3.

QUESTIONS AND PROBLEMS

2-1.—The kinematic viscosity of an oil is 0.021 ft^2/s and the specific gravity is 0.9. What is the absolute viscosity in the English, CGS, and SI systems and the kinematic viscosity in the CGS system?

2-2.—An oil tested by the Saybolt viscosimeter was found to have a viscosity of 150 SSU at room temperature. What is its kinematic viscosity in centistokes? In meters squared per second?

2-3.—Benzene has a kinematic viscosity of 8×10^{-5} ft^2/s. Determine its absolute viscosity in English and CGS units.

2-4.—Linseed oil has an absolute viscosity of 2.92×10^{-2} lb s/ft^2. Determine its kinematic viscosity in the English and SI systems.

2-5.—An oil has a viscosity of 225 SSU at 70° F. Determine its absolute and kinematic viscosities. The specific gravity of the oil is 0.92.

2-6.—A certain liquid has a specific weight of 56 lb/ft^3 and a kinematic viscosity of 100 centistokes at 70° F. Determine its absolute viscosity at this temperature.

2-7.—Water has an absolute viscosity of 1 centipoise at 68° F. Determine its kinematic viscosity at this temperature.

2-8.—Convert a kinematic viscosity of 0.001 ft^2/s to the corresponding value in the SI system.

chapter three

PROPERTIES PECULIAR TO INCOMPRESSIBLE FLUIDS

3-1 INTRODUCTION

Incompressible fluids (liquids) have several properties that gases do not have. Surface tension and vapor pressure are two of these which are put to use in industrial applications. Liquids also have the characteristic of not being completely incompressible. This tendency for some compressibility to exist is important enough in some applications that an allowance has to be made for it.

3-2 SURFACE TENSION

If a liquid has a free surface in contact with a gas such as the atmosphere, the free surface exhibits a *tension effect*. This tension effect, or *stressed skin*, is of sufficient magnitude that it will support small loads. For instance, a small needle placed gently in a horizontal position upon a water surface will not sink but will be supported by the tension in the liquid surface. Other examples that can be cited as resulting from the effects of surface tension are raindrops and soap bubbles. Figure 3-1 illustrates how surface tension acts on a droplet of a liquid. If the sphere in Fig. 3-1 is cut, the surface tension acting on the surface is as shown in part (b).

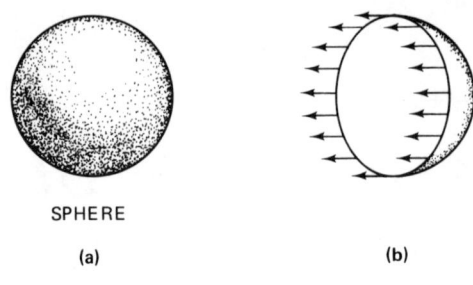

SPHERE

(a)   (b)

FIG. 3-1.

Surface tension also determines whether a liquid wets the surface it is on. Liquid mercury, when poured onto a surface such as glass, will not wet the surface. Instead, it stays in droplets, as shown in Fig. 3-2(a). Water will wet the glass, and the droplet of water in Fig. 3-2(b) breaks down and spreads out on the surface of the glass. The higher surface tension of the mercury (0.03562 lb/ft as compared to 0.004985 for water) causes this. Table 3-1 gives surface-tension values for some common liquids.

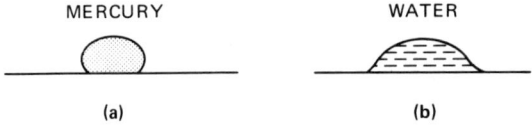

MERCURY   WATER

(a)   (b)

FIG. 3-2.

Table 3-1   **Surface Tension at 68° F in Contact with Air**

	lb/ft
Ethyl alcohol	0.001527
Carbon tetrachloride	0.001832
Turpentine	0.001857
Benzene	0.001980
Olive oil	0.002295
Water	0.004985
Mercury	0.03562

Surface tension also affects the junction of a liquid and a vertical solid surface. Figure 3-3 shows this effect for water and mercury against a vertical glass surface. The upward and downward radii are the result of the different surface-tension values of water and mercury.

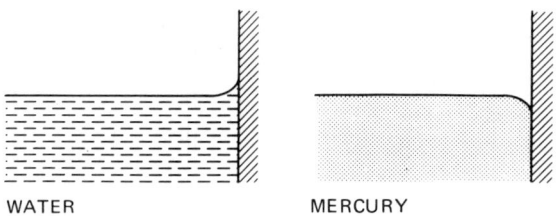

FIG. 3-3.

3-3 SURFACE TENSION AND CAPILLARITY

If a small tube (less than about $\frac{3}{8}$ in. inside diameter) is placed vertically in a liquid, the surface tension of the liquid causes capillary action to take place. *Capillary action (capillarity)* is illustrated in Fig. 3-4. Water rises in the tube, whereas mercury is depressed below the level of the liquid. The amount the liquid rises or falls depends on the size of the tube, with smaller-diameter tubes having the larger rise or fall. Above about $\frac{3}{8}$ in. inside tube diameter, capillarity does not occur.

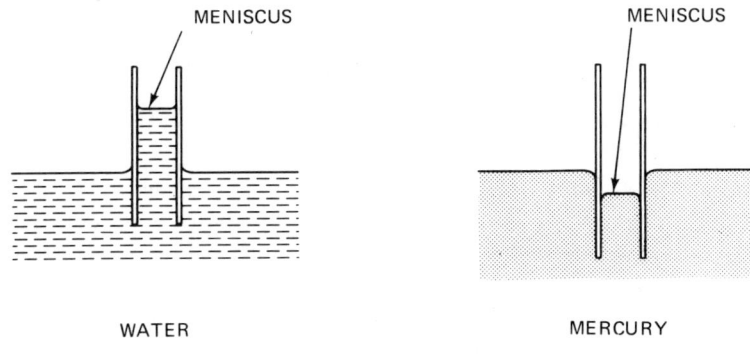

FIG. 3-4.

The curved surface formed in the tube is called a *meniscus*. Since small-bore glass tubes are frequently used in pressure-measuring gages (manometers), the curved meniscus tends to affect the accuracy of readings taken from the liquid level in the tube. Thus, when reading a gage, all readings should be taken at the level at the middle of the meniscus. Figure 3-5 illustrates this.

The field of instrumentation makes use of capillarity in such areas as supplying ink to a pen used to automatically record graphs, such as the one illustrated in Fig. 3-6.

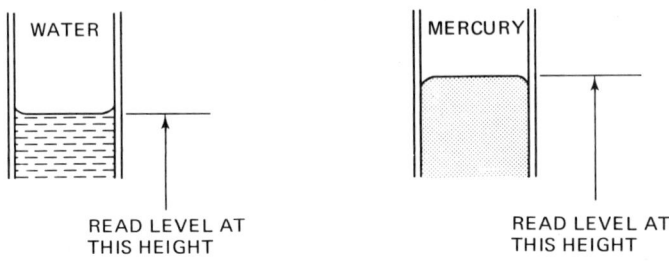

FIG. 3-5.

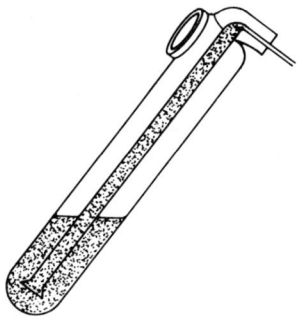

FIG. 3-6. Recorder pen with capillary-fed ink supply.

3-4 VAPOR PRESSURE

All liquids have a tendency to evaporate under normal atmospheric conditions. As an example, water evaporates and in the process goes from the liquid to the gaseous state. In the gaseous state it exerts a partial pressure called *vapor pressure*. An increase in temperature increases the rate of evaporation. Boiling—a rapid, violent form of evaporation—occurs when the temperature is high enough. In this case, the temperature must be high enough to raise the vapor pressure to a point equal to the atmospheric pressure. Thus, the vapor pressure of water at 212° F is 14.7 lb/in.^{2}, which is normal atmospheric pressure.

The rate of evaporation and the boiling point are also dependent on the pressure of the atmosphere (or gas) in which the liquid is contained. Water will boil at 200°F at a pressure of 11.5 lb/in.^{2}. Automobile cooling systems are pressurized to prevent loss of coolant due to boiling, which can occur at the lower atmospheric pressure found when traveling at a high elevation. A graph showing the relationship of the vapor pressure of water to temperature is shown in Fig. 3-7. Consideration of vapor pressure is important in designing any hydraulic system. If the pressure at any point in the system becomes low enough to equal the vapor pressure of the liquid at that temperature, vapor lock occurs, and liquid can no longer be pumped through the system. The vapor lock that sometimes occurs in an automobile engine on a hot summer day is an example of this. Table I in the Appendix lists values of vapor pressures of various liquids at specific temperatures.

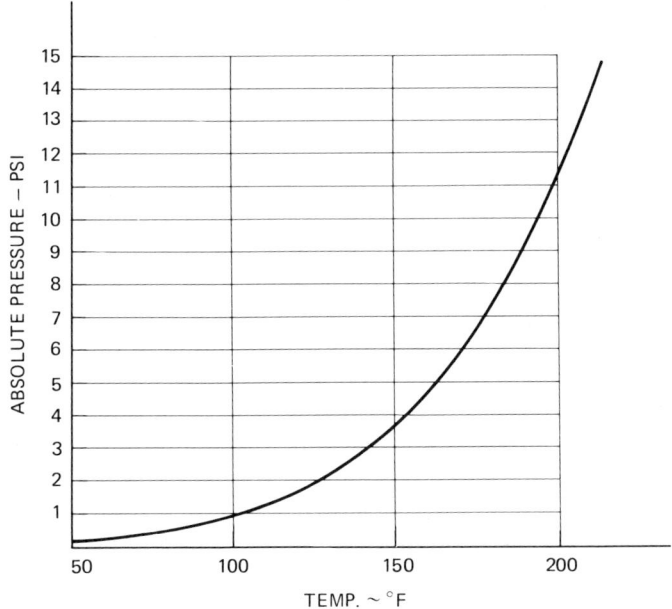

FIG. 3-7. Vapor pressure of water. (Data from *Steam Tables*, by Keenan/Keyes/Hill/Moore, 1969, *by permission of the publisher*, John Wiley & Sons, Inc.)

3-5   COMPRESSIBILITY OF LIQUIDS

The theoretical assumption that liquids are incompressible is accurate enough to use for most of the calculations and other work necessary to solve problems in hydraulics. Liquids actually are not completely incompressible, and the volume of a given quantity of liquid will decrease slightly if it is compressed.

The following guidelines may be used for computing the amount of reduction in volume for the liquids described:

      water: 0.33% reduction in volume for every 1000 lb/in.^{2} of pressure hydraulic oils: 0.5% reduction in volume for every 1000 lb/in.² of pressure

The following examples illustrate how these properties can be important in hydraulics.

EXAMPLE 1

The condensate (hot-water) pump shown in Fig. 3-8 pumps water from the condensate tank up to the storage tank. The temperature and pressure on the suction (low-pressure) side of the pump are 150° F and 4 psi. Is this safe to prevent vaporization of the water? (*Note:* Vaporization may cause cavitation, where pockets or cavities form in the liquid. Cavitation is potentially destructive to machine parts.)

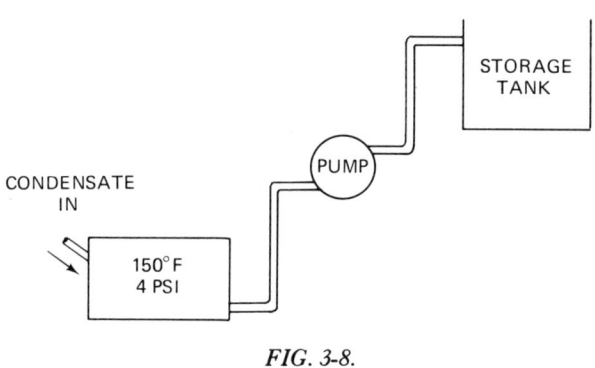

FIG. 3-8.

Solution

Refer to the graph of the vapor pressure of water, Fig. 3-7. The vapor pressure of water at 150° F is 3.7 psi. This is very close to the 4 psi in the condensate tank, and the system would not be safe from cavitation or vaporization of the water.

EXAMPLE 2

A hydraulic system operates at a pressure of 200 psi and has a capacity of 25 gal of hydraulic oil when not under pressure. Is the reduction in oil volume due to the pressure important enough to consider?

Solution

200 psi would only result in a reduction in volume of $\frac{200}{1000}$, or $\frac{1}{5}$, of 0.5% of the original volume (we have seen that 1000 psi will reduce

the volume 0.5%). This quantity is so small that it normally would be neglected.

QUESTIONS AND PROBLEMS

3-1.—Determine the vapor pressure of water at 100° F.

3-2.—A hydraulic system operates at a nominal 2500 psi pressure. Its nonpressurized volume is 45 gal of hydraulic oil. Determine the amount that the volume of oil is reduced under this pressure.

3-3.—In which one of the capillary tubes in Fig. 3-9 will water rise the highest?

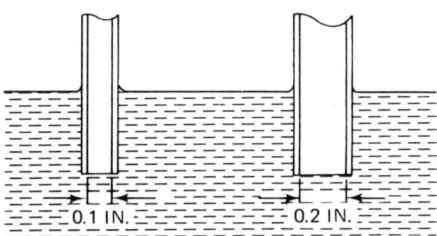

FIG. 3-9.

3-4.—At what atmospheric pressure would water boil at 160° F?

3-5.—Which of these liquids evaporates faster at 68° F—gasoline, ethyl alcohol, or kerosene? Explain how you arrived at your answer.

chapter four
COMPRESSIBLE FLUIDS

4-1   INTRODUCTION

By the definitions of Chapter One, the compressible fluids are gases. They share some of the characteristics of liquids but are different in behavior in many other ways. By far the most common gas used in practical applications in the fluid power industry is air. Whereas "hydraulics" is the common term used to describe the use of liquids in fluid power, the term "pneumatics" is frequently used where air is the fluid used.

In some cases gases other than air are used. Since air is more readily available and is cheaper, some special requirement in the application would dictate the use of another gas.

4-2   PRESSURE MEASUREMENT

It is necessary to define the physical basis for the measurement of pressure when dealing with gases. These definitions also apply in all cases to the measurement of pressures when liquids are used.

Development of a method to measure the pressure of the earth's atmosphere was necessary before the theory and application of pressure measurement could be developed. In 1643 Evangelista Torricelli, an Italian scientist, developed the mercury manometer (Fig. 4-1) for this purpose. In the *barometer* (the manometer or similar gage, when used to measure atmospheric pressure, is called a barometer) a closed tube is filled with mercury while in the upright position. It is then inverted in an open container filled with mercury. The mercury in the tube falls until it stops at a height h above the level of mercury in the dish. The pressure P_2 now equals zero (it is actually the vapor pressure of mercury, which is so small it can be neglected). Atmospheric pressure acts on the open mercury in the dish and supports the column of mercury in the tube.

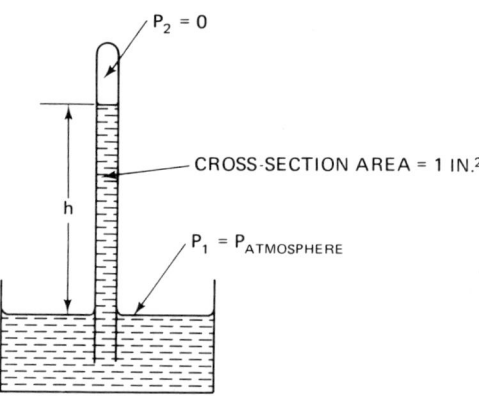

FIG. 4-1.   Torricelli's barometer.

Since atmospheric pressure supports the column of mercury of height h, h is proportional to the atmospheric pressure. This fact was originally proved by experiments conducted at different elevations. The experiments showed that the value of h is lower at higher elevations and that lower pressure exists at higher elevations.

The basic pressure units are force per unit area. If the column cross-sectional area is 1 in.^{2}, as shown in Fig. 4-1, the volume is $h \times$ (1 in.^{2}). The weight of the column is then the specific weight, W_s, of mercury multiplied by the volume:

$$\text{weight} = W_s \times (\text{vol.})$$

The specific weight of mercury from Table I in the Appendix is 848.6 lb/ft^3. Substitution yields

$$\text{weight} = \frac{848.6 \text{ lb/ft}^3}{1728 \text{ in.}^3/\text{ft}^3} (h \text{ in.}^3)$$

The value of h has been found to be 76 cm, or 29.92 in., under 1 standard

Sect. 4-2 / PRESSURE MEASUREMENT

gravity. Using this,

$$\text{weight} = \frac{848.6(29.92)}{1728} \text{ lb}$$

$$= 14.7 \text{ lb}$$

The weight or force can be equated to pressure since it acts on 1 in.^{2} of area, so the pressure is

$$P = \frac{F}{A} = \frac{14.7 \text{ lb}}{1 \text{ in.}^2} = 14.7 \text{ lb/in.}^2$$

Thus, the value of the standard atmosphere is 14.7 lb/in.^{2}.

Both the standard atmosphere and zero pressure, as at P_2, are used as reference points for pressure measurement. Zero pressure is called *absolute pressure* in practice. The range between zero pressure and 1 atm is a vacuum.

Most pressure gages measure pressure with reference to the atmosphere. Measured pressure referenced to atmosphere is called *gage pressure*. Positive gage pressure is greater than 1 atm. Negative gage pressure is pressure less than 1 atm and is a *vacuum*.

The relationship between absolute and gage pressure is shown in Fig. 4-2. In the English system the most common unit of pressure is the pound per square inch, abbreviated lb/in.^{2} or psi. Gage pressure is referred to as pound per square inch gage, abbreviated *psig*. Absolute pressure is referred to as pound per square inch absolute, abbreviated *psia*.

The term inches of mercury vacuum (in. Hg vacuum) is shown in Fig. 4-2. In this method of measuring a vacuum, atmospheric pressure is

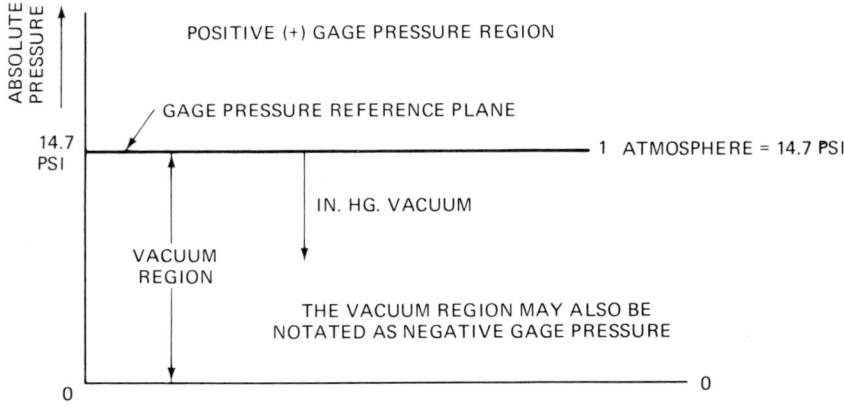

FIG. 4-2. Absolute and gage pressure.

used as the reference. Thus, 10 in. Hg vacuum means a pressure equivalent to 10 in. Hg below atmospheric pressure.

The pressure unit in the SI system is the pascal (Pa), equal to 1 newton per square meter. Table II in the Appendix has conversion units to change from one system to the other.

4-3 TEMPERATURE

Like pressure, temperature, when used in calculations for gases, requires the use of an absolute-temperature scale. The absolute temperature has been defined as the temperature at which all molecular activity of the gas ceases. The temperature scale used in the English system is the *Rankine* (R) temperature scale. The corresponding SI system absolute-temperature scale is the *Kelvin* scale. Use the following equations to convert from degrees Fahrenheit (°F) and degrees Celsius (°C) to these scales:

$$\text{degrees Rankine (°R)} = \text{degrees Fahrenheit (°F)} + 460 \qquad (4\text{-}1)$$

$$\text{degrees Kelvin (K)} = \text{degrees Celsius (°C)} + 273 \qquad (4\text{-}2)$$

Some equivalent temperature points on all four scales are shown in Fig. 4-3.

Equation (4-2) contains a peculiarity of nomenclature. The degree symbol (°) has been associated with temperature scales over the years; 100 degree Fahrenheit has been written 100° F, for example. However, in the SI system degrees Kelvin is written K, without the degree symbol. De-

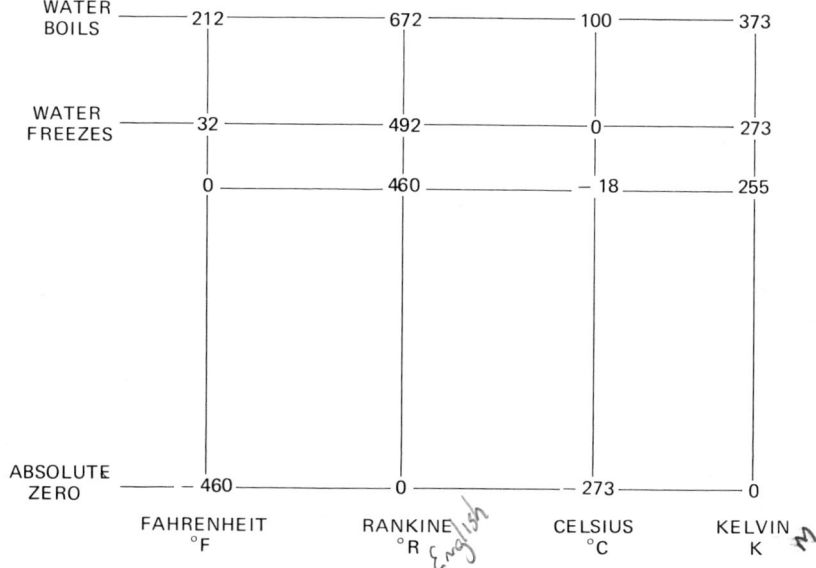

FIG. 4-3. Temperature scales.

grees Celsius does use the degree symbol, to avoid confusion with an electrical unit, C (coulomb). (We should also note that the Celsius scale was previously called the Centigrade scale.)

EXAMPLE 1

The output side of an air compressor has a gage on it which reads 95 psi. Is this reading absolute or gage pressure?

Solution

Since a gage is referenced to atmospheric pressure, the pressure read on the instrument is gage pressure.

EXAMPLE 2

What is the absolute pressure corresponding to 95 psig?

Solution

Referring to Fig. 4-2, the 95 psig is referenced to atmospheric pressure of 14.7 psi. The absolute pressure is then the sum (algebraic) of the two:

$$\text{Absolute pressure} = 14.7 + 95 = 109.7 \text{ psia}$$

EXAMPLE 3

The gage pressure on a certain system is −3.2 psi. What is the absolute pressure?

Solution

Referring again to Fig. 4-2, note that a negative gage pressure would fall in the vacuum region and that mathematically it can be referred to as negative. Keeping the signs, the absolute pressure can be determined by algebraically adding, as follows:

$$\text{Absolute pressure} = 14.7 + (-3.2) = 14.7 - 3.2 = 11.5 \text{ psia}$$

EXAMPLE 4

Convert a temperature of 122° F to °R.

Solution

Using Eq. (4-1),

$$°R = °F + 460$$
$$= 122 + 460 = 582° R$$

EXAMPLE 5

Convert a temperature of $-10°$ C to K.

Solution

Using Eq. (4-2),

$$K = °C + 273$$
$$= -10 + 273 = 263 \text{ K}$$

4-4 KINETIC THEORY OF GASES

The kinetic theory of gases has been developed from the work done by physicists from the seventeenth century to the present. Some of the assumptions of this theory are:

1 / The gas consists of a large number of particles called *molecules*.

2 / Molecules are constantly in random motion at different velocities.

3 / Collisions of molecules result in an internal energy of the gas.

4 / Temperature and pressure affect the molecule activity and the internal energy.

Figure 4-4 illustrates molecular activity of this type. The work of these physicists led to the formulation of the gas laws, which are described in the following pages.

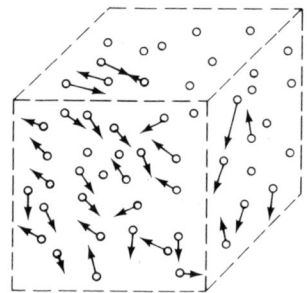

FIG. 4-4.   Kinetic energy of gas molecules.

4-5 BOYLE'S LAW

In the seventeenth century Charles Boyle, an Irish scientist, showed by experiment that the volume of a given quantity varies inversely with absolute pressure if the temperature is held constant. Stated mathematically, this is

$$\frac{P_1}{P_2} = \frac{V_2}{V_1}$$

and

$$P_1 V_1 = P_2 V_2 \qquad (4\text{-}3)$$

where the subscripts 1 and 2 indicate the different pressure and volume conditions. Since any other condition (P_3, V_3) at the same temperature would also equal $P_1 V_1$, it follows that, under constant temperature, $P_1 V_1 = P_2 V_2 = P_3 V_3 =$ constant. Therefore,

$$PV = k \qquad (4\text{-}4)$$

where k is a constant.

4-6   CHARLES' LAW

Charles' law, named for the French scientist Jacques Charles, states the following:

> **Charles' law:** If the volume of a given mass of gas is held constant, the absolute pressure varies directly with the absolute temperature.

Using $k =$ a constant,

$$P = kT \qquad (4\text{-}5)$$

If different conditions are used, as indicated by the subscripts 1 and 2, two equations can be written:

$$P_1 = kT_1 \quad \text{and} \quad P_2 = kT_2$$

Dividing one by the other,

$$\frac{P_1}{P_2} = \frac{kT_1}{kT_2}$$

The constant k drops out, giving

$$\frac{P_1}{P_2} = \frac{T_1}{T_2} \qquad (4\text{-}6)$$

Equation (4-6) is another form of Charles' law.

4-7 PERFECT-GAS LAW

A third gas law is developed from Boyle's law and Charles' law. When a gas is considered under three different conditions involving temperature, pressure, and volume, the perfect-gas law is developed.

The *perfect-gas law* states that a perfect gas follows the following mathematical relationship:

$$PV_s = RT \qquad (4\text{-}7)$$

where P = absolute pressure
V_s = specific volume of the gas
R = a constant, the value of which depends on the gas
T = absolute temperature, °R

In other words, the absolute pressure of a gas multiplied by its specific volume equals the gas constant R multiplied by the temperature.

Another form of this same equation can be obtained by using the total weight W of the gas and total volume V. If both sides of Eq. (4-7) are multiplied by W, the equation becomes $PV_sW = WRT$. But $V_sW = V$, the total volume of the gas, and the equation then becomes

$$PV = WRT \qquad (4\text{-}8)$$

No gas follows this law perfectly. However, most gases follow the law closely enough so that the equation can be used satisfactorily.

The gas constant, R, varies for different gases, and Table IV in the Appendix lists values of R for some common gases. Equation (4-7) or its other form, Eq. (4-8), are sometimes referred to as the *equation of state for a perfect gas*.

4-8 UNIVERSAL GAS CONSTANT

A quantity called the *universal gas constant* is obtained by multiplying the molecular weight m of the gas by R. The molecular weight of a compound is defined as the sum of the atomic weight of all the atoms in its chemical formula. Molecular weights of some of the common gases are listed in Table IV of the Appendix.

The quantity mR is nearly constant for all gases that follow the perfect-gas law. Its value in the English system of units is 1545 ft lb/°R. Equation (4-8) can be used in still another form by substituting the mR term in place of WR. It then becomes

$$PV = mRT \qquad (4\text{-}9)$$

V would then equal the volume occupied by a molecular weight (1 mole) of gas.

4-9 UNITS OF THE PERFECT-GAS LAW

Engineering practice in the United States is to use primarily the English system of units for calculating gas volumes, pressures, and temperatures. In order to make the units consistent in Eqs. (4-7) and (4-8), pressures should be in lb/ft² and volumes in ft³. The units of R are ft lb per lb per °R. The units of Eq. (4-7) then are:

$$\frac{\text{lb}}{\text{ft}^2} \frac{\text{ft}^3}{\text{lb}} = \frac{\text{ft lb}}{\text{lb °R}}$$

where P is in lb/ft² absolute
    V_s is in ft³/lb

The units of Eq. (4-8) are:

$$\left(\frac{\text{lb}}{\text{ft}^2}\right) \text{ft}^3 = \text{lb} \left(\frac{\text{ft}}{\text{°R}}\right) \text{°R}$$

where P is in lb/ft²
    V is in ft³ (total volume)
    W is in lb (total weight)

In the SI system the standard units for energy, kilogram, and temperature can be used in both the equation for the perfect-gas law and for R, the constant. The mR term is

$$mR = 8.314 \text{ joules per gram mole per degree Kelvin}$$
$$= 8.314 \frac{J}{\text{g-mol K}} \quad (4\text{-}10)$$

The PV term of the equation $PV = mRT$ would be in joules per gram mole, to be consistent with Eq. (4-10).

Since 1 joule = 1 newton meter (N·m), the newton meter is substituted to obtain the required units of force length/mass, which result in the PV term.

4-10 DISCUSSION AND PROBLEMS

In working problems involving compressible fluids it is well to remember that the pressure, temperature, and volume may all vary at the same time. When this occurs, pressure, temperature, and volume values at one set of conditions cannot be used at another set of conditions. The one thing that does remain constant is the mass of the gas. It is therefore frequently necessary to determine the mass (the weight will actually be used instead of mass in the calculations) of the gas.

EXAMPLE 1

A compressor supplies air at 100 psi to tank 1 as shown in Fig. 4-5. Ten cubic feet of the air at the conditions of tank 1 are transferred to tank 2. The pressure in tank 2 is regulated to 20 psi by the regulator. What is the temperature of the air in tank 2?

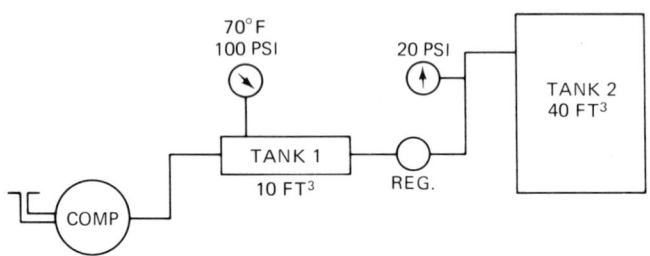

FIG. 4-5.

Solution

The perfect-gas law can be used to solve the problem. It is necessary to consider the weight of air transferred from tank 1 to tank 2 even if the actual weight is not determined. Using Eq. (4-8) and letting subscripts 1 and 2 indicate the conditions in tank 1 and tank 2, respectively,

$$P_1 V_1 = WRT_1 \quad \text{and} \quad P_2 V_2 = WRT_2$$

A ratio can be set up by dividing the first equation by the second, as follows:

$$\frac{P_1 V_1}{P_2 V_2} = \frac{WRT_1}{WRT_2}$$

The *WR* terms cancel and there is then

$$\frac{P_1 V_1}{P_2 V_2} = \frac{T_1}{T_2}$$

Solving for T_2,

$$\frac{T_2}{T_1} = \frac{P_2 V_2}{P_1 V_1} \quad T_2 = \frac{T_1 P_2 V_2}{P_1 V_1}$$

The values to be used in these equations are:

$$T_1 = 70 + 460 = 530° \text{R}$$

$$P_1 = (100 + 14.7)\,\frac{\text{lb}}{\text{in.}^2}\left(144\,\frac{\text{in.}^2}{\text{ft}^2}\right) = 16{,}520\,\frac{\text{lb}}{\text{ft}^2}$$

Sect. 4-10 / DISCUSSION AND PROBLEMS 35

$$P_2 = (20 + 14.7) \frac{\text{lb}}{\text{in.}^2} \left(144 \frac{\text{in.}^2}{\text{ft}^2}\right) = 5000 \frac{\text{lb}}{\text{ft}^2}$$

$$V_1 = 10 \text{ ft}^3$$
$$V_2 = 40 \text{ ft}^3$$
$$R = 53.3 \frac{\text{ft lb}}{\text{lb °R}} \qquad \text{(from Table IV in the Appendix)}$$

Note that the pressure and temperature values have been converted to absolute. Substituting the values in the equation,

$$T_2 = \frac{(530)(5000)(40)}{(16{,}520)(10)} = 642° \text{ R}$$

or

$$T_2 = 642 - 460 = 182° \text{ F}$$

EXAMPLE 2

The molecular weight of oxygen is 32 (see Table IV in the Appendix). What is the volume of 1 lb-molecular weight of oxygen at 14.7 psia and 460° R?

(*Note:* The molecular weight expressed in pounds is called a *mole*. The corresponding value in the metric system is expressed in grams. Thus, the problem is to find what volume 32 lb of oxygen occupies under the conditions stated.)

Solution

The answer can be obtained by using Eq. (4-9) and the universal gas constant. These are $PV = mRT$ and the value of 1545 lb ft/°R. Pressure and temperature values are already in absolute units and can be used as they are except that pressure has to be converted to lb/ft². Then the equation becomes

$$V = \frac{mRT}{P}$$

where $mR = 1545 \dfrac{\text{lb ft}}{\text{°R}}$

$T = 460° \text{ R}$

$P = \left(14.7 \dfrac{\text{lb}}{\text{in.}^2}\right)\left(144 \dfrac{\text{in.}^2}{\text{ft}^2}\right) = 2117 \dfrac{\text{lb}}{\text{ft}^2}$

and

$$V = \frac{(1545)(460)}{2117} = 336 \text{ ft}^3$$

A more exact answer for some gases could be obtained by using the *m* and *R* values from Table IV in the Appendix, multiplying them, and using the quantity instead of 1545. This is because the true value of *R* is listed in Table IV. The universal gas constant of 1545 lb ft/°R is based on a perfect gas, and there is a small error in using it when the gas under consideration departs from perfect-gas characteristics.

EXAMPLE 3

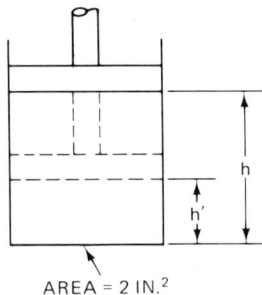

A cylinder with the bottom end closed and the top end open to the atmosphere has a close-fitting piston inserted as shown in Fig. 4-6. Assuming no friction and no leakage of air, how far down will the piston drop? The piston weight is 20 lb.

FIG. 4-6.

Solution

The initial pressure is atmospheric and constant-temperature conditions can be assumed, since there is no heat added or taken away during the process and no temperatures are given in the problem. Then Boyle's law applies and

$$P_1 V_1 = P_2 V_2 \qquad (4\text{-}3)$$

Solving for V_2,

$$V_2 = \frac{P_1 V_1}{P_2}$$

The pressure the air is under when the piston comes to rest is 20 lb/2 in.² = 10 psig. The height h and h' multiplied by the constant area of 2 in.² equals the volume at the two conditions. Substituting these values,

$$2h' = \frac{(14.7)(2h)}{10 + 14.7}$$

$$h' = 0.595h$$

The distance the piston drops is $h - h'$, or

$$\text{distance} = h - 0.595h = 0.405h$$

Note that the absolute pressures were used in the equation without converting to lb/ft^2, since the units cancel in the equation.

QUESTIONS AND PROBLEMS

4-1.— 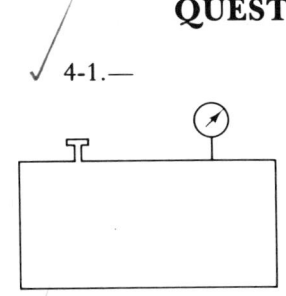 The closed tank has a pressure gage attached to it to sense the pressure inside the tank in Fig. 4-7. The gage shows a reading of 0 psi. What is the absolute pressure?

FIG. 4-7.

4-2.—Determine the specific volume and specific weight of air at 14.6 psia and 85° F.

4-3.—Ten cubic feet of nitrogen under constant pressure is heated from 90° to 350° F. What is its final volume?

4-4.—Convert the following pressures to absolute.
    a— -2.1 psig b—20 psig
    c—1 psig d—0 psig

4-5.—Convert the following temperatures.
    a— $-10°$ F to Rankine b—110° F to Rankine
    c—0° C to Kelvin d—120° C to Kelvin

4-6.—Valve 1 is shut between the two tanks in Fig. 4-8. Tank 1 contains 10 ft^3 of nitrogen under 100 psig pressure. Tank 2 is

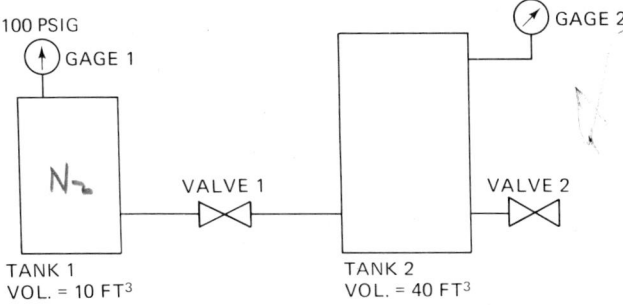

FIG. 4-8.

evacuated by vacuum pump through valve 2 so that its absolute pressure is zero. Valve 2 is then shut and valve 1 is then opened. What will be the reading on gage 2? The temperature remains constant.

4-7.—How many pounds of air are required to fill a cylinder 1 ft long and 1 ft in diameter to a pressure of 30 psig and temperature of 70° F?

4-8.—What volume will 1 lb of air occupy at 25 psig pressure and 75° F?

4-9.—What volume will 1 lb of air occupy at −4 psig pressure and 50° F?

4-10.—Fifteen cubic ft of air at the conditions of tank 1 in Fig. 4-9 is transferred to tank 2 under constant-temperature conditions. What is the pressure in tank 2? The air in tank 2 has been pumped out previously so that its pressure is 29.92 in. Hg vacuum.

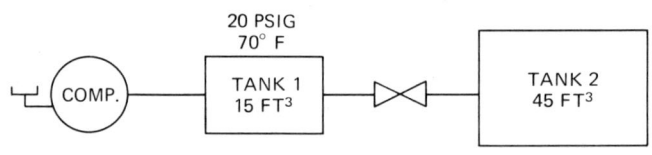

FIG. 4-9.

4-11.—One gram mole of a certain gas occupies a volume of 100 liters at a temperature of 100 K. What is the pressure in pascal units?

chapter five

THERMODYNAMIC PROCESSES OF COMPRESSIBLE FLUIDS

5-1　INTRODUCTION

When studying compressible-fluid behavior it is necessary to touch upon the thermodynamic processes that govern compressible fluids. Thermodynamics is a science dealing with energy and energy transformations occurring as a result of a change that takes place in a substance or medium. As its name indicates, it is related to heat energy, as differentiated from other forms of energy, such as mechanical or electrical energy. The compression of a gas is an example of thermodynamics. It requires energy in the form of mechanical work to perform the compression. In the process the volume of the gas and also its pressure changes. When compressed, the gas has stored internal energy that can be used to do mechanical work. Thus, the study of fluid mechanics is interrelated with the study of thermodynamics.

This is illustrated by Fig. 5-1. Here there is a small gasoline engine which powers an air compressor. The air compressor compresses air and it is stored under pressure in a tank of 20 ft^3 volume and at 100 psi pressure. When used, the air is allowed to flow from the tank into the double-acting cylinder, which does mechanical work of some sort. The heat energy of thermodynamics occurs or exists in the following ways in the cycle of Fig. 5-1:

fuel combustion   mechanical frictional losses
gas compression   fluid frictional losses

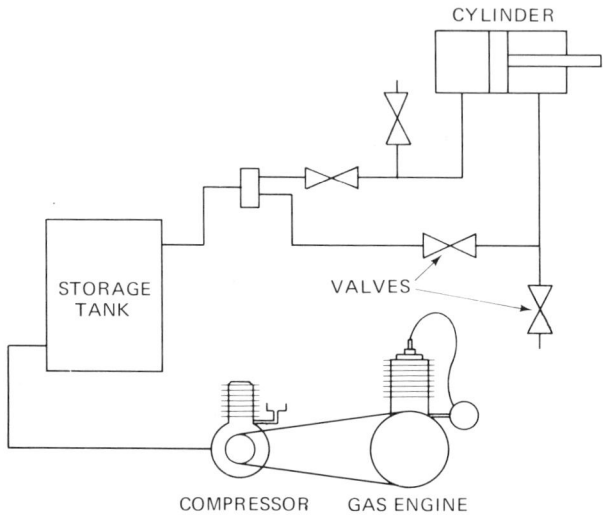

COMPONENT(S)	ENERGY TRANSFORMATION
GAS ENGINE	FUEL COMBUSTION → GAS EXPANSION → MECH. ENERGY
ENGINE – COMPRESSOR	MECH. → MECH.
COMPRESSOR	MECH. → GAS COMPRESSION → STORED ENERGY
CYLINDER	GAS EXPANSION → MECH. ENERGY

FIG. 5-1.

5-2   *ENERGY UNITS*

All the forms of energy that will be discussed here can be described by common units. The common unit of work energy in the English system is the foot pound. It can be defined as the product of a force acting on a particle multiplied by the distance the particle is caused to move by the force. The distance of movement of the particle must be in a straight line and in the same direction of the force. The standard energy unit in the SI system is the joule, equal to the newton meter. Since the newton is a unit of force and the meter a unit of length, the same basic physical quantities are provided that exist in the English system.

In heat energy the common form of energy unit is the British thermal unit (Btu). It is defined as the amount of heat required to raise the temperature of 1 lb of water 1° F, when the water is at a temperature of 63° F. The joule, however, is also used as the basic unit for heat energy in the SI system.

To convert heat energy to work energy, the following relationships apply:

$$1 \text{ Btu} = 778 \text{ ft lb} \qquad (5\text{-}1)$$
$$1 \text{ Btu} = 1055 \text{ joules}$$

5-3 ISOTHERMAL PROCESS

The word "isothermal" in itself means constant temperature. An *isothermal process* is therefore a process in which the temperature remains constant. The condition described by Boyle's law is an isothermal one and the equation for the condition states this mathematically:

$$P_1 V_1 = \text{constant} = P_2 V_2 \ldots \qquad (5\text{-}2)$$

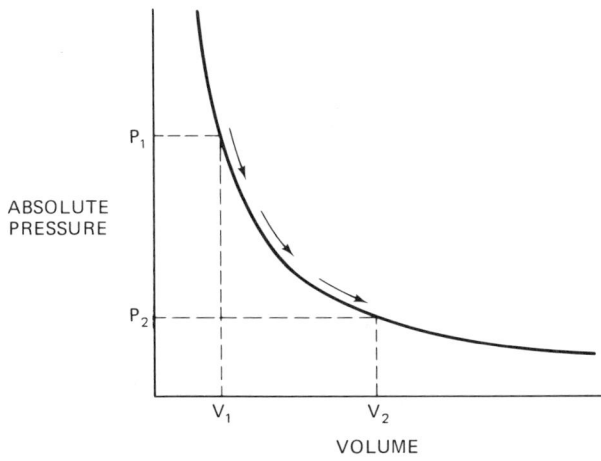

FIG. 5-2. Pressure versus volume for the isothermal process.

A graphical plot of pressure versus volume is a help in analyzing the constant-temperature process. Such a plot is shown in Fig. 5-2. This process could occur in a cylinder as shown in Fig. 5-3. The initial volume is V_1 and the final volume is V_2. The pressure drops from P_1 to P_2 and the path of the process is shown on the curve of Fig. 5-2. We should note that heat would have to be added to the gas to keep the gas temperature constant. Equation (5-2) is sometimes written in another form to allow direct comparison with the equations for the adiabatic and polytropic

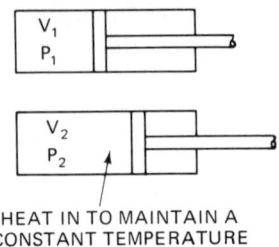

HEAT IN TO MAINTAIN A
CONSTANT TEMPERATURE

FIG. 5-3. Expansion in the isothermal process.

processes, which are described next. This other form is

$$P_1 V_1^n = P_2 V_2^n \qquad (5\text{-}2)$$

where $n = 1$.

5-4 ADIABATIC PROCESS

By definition there is no heat transfer during the adiabatic process. Thus, the adiabatic process is one in which no heat energy enters or leaves the working medium during the change that takes place. The same cylinder as shown in Fig. 5-3 may be used to demonstrate the adiabatic process as long as there is no heat transfer into or out of the fluid. The pressure-volume relationship varies from that of the isothermal process and is shown in Fig. 5-4. The equation for the adiabatic process is the same as Eq. (5-2) except that the value of n changes. The term n becomes known as the *adiabatic exponent* and its value varies with the gas. For air the value of the

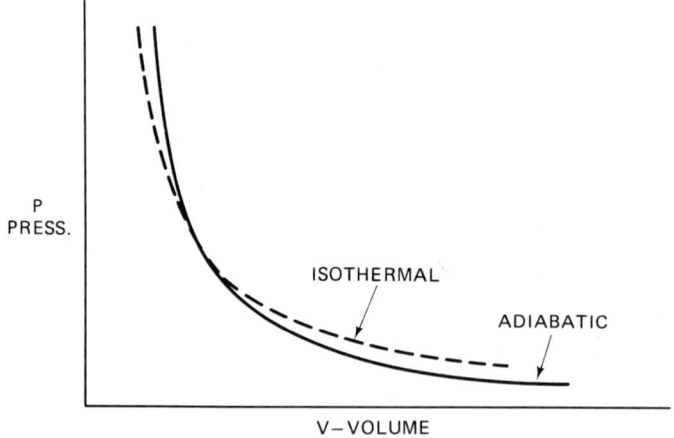

FIG. 5-4. Pressure vs. volume for the adiabatic process and the isothermal process.

adiabatic exponent is 1.4. Values for other common gases are listed in Table IV in the Appendix. When air is the medium and the process can be considered adiabatic, Eq. (5-2) then becomes

$$\text{air} \longrightarrow \quad P_1 V_1^{1.4} = P_2 V_2^{1.4} = \text{constant} \quad (5\text{-}3)$$

Processes are sometimes assumed to be adiabatic for ease in analyzing them. In practice some heat transfer almost always occurs.

5-5 POLYTROPIC PROCESS

To overcome the disadvantage associated with the assumption of the adiabatic process, the value of n is changed to represent true conditions more closely in the *polytropic process*. Its value must be determined experimentally for each gas in this case. The polytropic process, which closely represents the actual conditions that occur when air is compressed, has a typical value of 1.3 for n. However, no general value of n can be given, so the equation remains in general terms:

$$P_1 V_1^n = P_2 V_2^n \quad (5\text{-}4)$$

5-6 REVIEW

The equations for the three processes that have been discussed are now repeated:

Isothermal: $\quad P_1 V_1^n = P_2 V_2^n \quad\quad n = 1$

Adiabatic: $\quad P_1 V_1^n = P_2 V_2^n \quad\quad n$ varies with the gas; Table IV in the Appendix lists the values

Polytropic: $\quad P_1 V_1^n = P_2 V_2^n \quad\quad n$ determined experimentally for each gas

For air motors and compressors operating in the range of several hundred pounds pressure, the polytropic process is the closest to the actual process.

EXAMPLE 1

Air is compressed adiabatically from an initial pressure of 24 psig to a final volume of one half its original volume. Determine its final pressure.

Solution

The compression is adiabatic, so Eq. (5-3) applies:

$$P_1 V_1^{1.4} = P_2 V_2^{1.4}$$

but
$$P_1 = (23 + 14.7)\frac{\text{lb}}{\text{in.}^2}\left(144\frac{\text{in.}^2}{\text{ft}^2}\right) = 5573\frac{\text{lb}}{\text{ft}^2}$$

$$V_2 = \tfrac{1}{2}V_1 \quad \text{and} \quad V_1 = 2V_2$$

Substituting and solving for P_2:

$$(5573)(2V_2)^{1.4} = P_2 V_2^{1.4}$$

$$P_2 = \frac{(5573)(2V_2)^{1.4}}{V_2^{1.4}} = (5573)(2)^{1.4}$$

$$= (5573)(2.64) = 14{,}713\frac{\text{lb}}{\text{ft}^2}$$

$$= \frac{14{,}713}{144}\frac{\text{lb}}{\text{in.}^2} = 102\frac{\text{lb}}{\text{in.}^2} \text{ absolute}$$

Note that the conversion from lb/in.² to lb/ft² could have been omitted in the first step, since the pressure terms on both sides of the equation are identical. If it had been omitted, the final pressure P_2 would have been in lb/in.² instead of lb/ft².

EXAMPLE 2

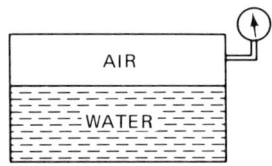

The pressurized water tank in Fig. 5-5 contains 150 gal of water under 35 psig pressure. The total tank volume is 200 gal. Fifty gal of water is drawn off. Assuming isothermal conditions, what will be the pressure in the tank?

FIG. 5-5.

Solution

For isothermal conditions, $P_1 V_1 = P_2 V_2$. The initial volume of air $V_1 = 200 - 150 = 50$ gal s.

$$V_1 = \frac{50 \text{ gal}}{7.48 \text{ gal/ft}^3} = 6.68 \text{ ft}^3$$

$$P_1 = (35 + 14.7)\frac{\text{lb}}{\text{in.}^2} \quad \text{(the units are consistent, so they are left in psi)}$$

$$V_2 = 200 - 100 = 100 \text{ gal}$$

$$= \frac{100 \text{ gal}}{7.48 \text{ gal/ft}^3} = 13.4 \text{ ft}^3$$

$$P_2 = \frac{P_1 V_1}{V_2} = \frac{(49.7)(6.68)}{13.4} = 24.8 \text{ psia}$$

QUESTIONS AND PROBLEMS

5-1.—Twenty-five cubic feet of air at atmospheric pressure is compressed to a volume of 10 ft.^{3} Compute the final pressure under:
  a—Isothermal conditions.
  b—Adiabatic conditions.
  c—Polytropic conditions.
  Use $n = 1.3$.

5-2.—The accumulator is charged with nitrogen under 1000 psig pressure in the bladder as shown in Fig. 5-6. The initial volume of the bladder at this pressure is 1.3 ft^3. Hydraulic fluid at 1500 psig then flows through the valve and the pressure stabilizes at 1500 psig. Assuming adiabatic conditions, what is the final volume of the bladder?

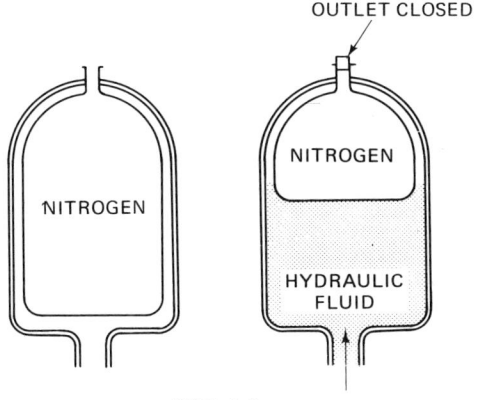

FIG. 5-6.

5-3.—In 1 hour a gasoline engine puts out 123,440 ft lb of work. Convert this value to heat energy in Btu.

5-4.—Five pounds of air is expanded isothermally from 100 psig and 70° F to atmospheric pressure. Determine the final volume.

5-5.—Ten pounds of nitrogen is expanded adiabatically from 250 psia and 75° F to atmospheric pressure. Determine the initial and final volumes.

5-6.—Carbon dioxide is compressed from atmospheric pressure and 70° F to a pressure of 200 psia. The initial volume is 27 ft^3. Determine the final volume if
  a—The compression is isothermal.
  b—The compression is adiabatic.

5-7.—Ten cubic feet of nitrogen at atmospheric pressure and 75° F is expanded adiabatically by reducing the pressure to 2 psi less than atmospheric. Determine the final volume.

5-8.—A certain gas follows the polytropic process governed by the term $PV^{1.2}$. If the initial pressure of 1500 m³ is 122 kilopascals absolute, what is the volume when the pressure is increased to 250 kilopascals absolute?

chapter six

FLUID STATICS

6-1 INTRODUCTION

Fluids in use are always under one of two conditions, static or dynamic. The static condition is defined as one in which there is no relative motion between fluid particles. Since there is no relative motion between fluid particles, viscosity can have no effect on the fluid behavior. This makes the analysis of fluid statics somewhat simpler than fluid dynamics. Conversely, in fluid dynamics there is relative motion between fluid particles, and viscosity does have an effect.

6-2 PRESSURE–HEIGHT RELATIONSHIPS

Pressure has previously been defined as force per unit area and the concept of a column of liquid with a uniform cross-sectional area has been used to illustrate it. This column of liquid is shown again in Fig. 6-1. With the uniform cross-sectional area of the column, and assuming uniform density of the fluid from top to bottom, it has been shown that the pressure is proportional to the height of the column. This can be stated as follows:

Pressure–height relationship: Under static conditions and with no other internal pressure, the pressure at any point in a fluid system is proportional to the height of the fluid above that point.

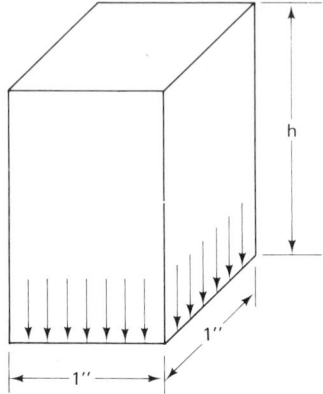

FIG. 6-1. Column of liquid exerting a pressure P on the bottom of a container.

This variation of pressure with height or elevation was brought out earlier in the discussion of Pascal's law. The meaning of the statement above is further explained in Fig. 6-2. Here there are two vertical lengths of pipe which have been filled with a fluid. The pipe in Fig. 6-2(a) has been filled

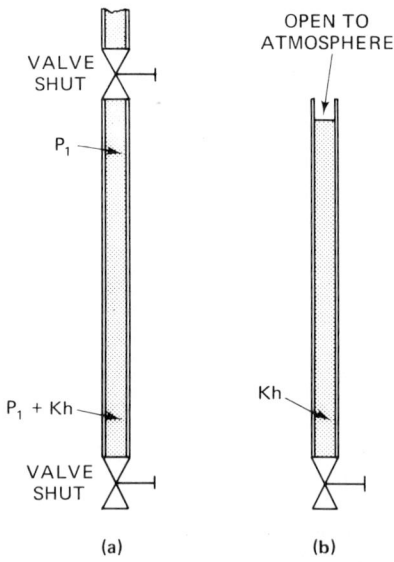

FIG. 6-2. K = constant of proportionality.

with a fluid, internally pressurized to a pressure of P_1, and the valves closed on both ends. The pressure at the top of the pipe is P_1 and the pressure at the bottom is $P_1 + kh$, where k is a constant of proportionality which depends on the fluid density.

In Fig. 6-2(b) the pipe is closed at the bottom but open to the atmosphere. In this case the pressure at the bottom of the pipe is equal to kh. The pressure at the top of the pipe is atmospheric.

The constant of proportionality k is used here for convenience, and values that are used for k are covered later. Note that, in Fig. 6-2, all the pressure measurements are with respect to atmosphere and are thus gage pressures.

6-3 PRESSURE HEAD

The height of a column of fluid can be used as a pressure-measuring device. The manometer is a good example of such an instrument. Since the density of the fluid has to be of significant magnitude, liquids are used instead of gases.

When pressure is caused or measured by a column of liquid, the term "pressure head" is frequently used. In common usage, "head" and "pressure" have come to mean the same thing.

The concept of pressure head as the equivalent of a column of liquid can be explained further by the manometer in Fig. 6-3, a mercury manometer used to measure the pressure in a vertical pipe filled with water. Water is flowing in the pipe with a certain velocity and under a certain pressure. This pressure has displaced the mercury in the manometer column upward to a distance of h. If the water were not flowing and instead were under a static pressure, the manometer would

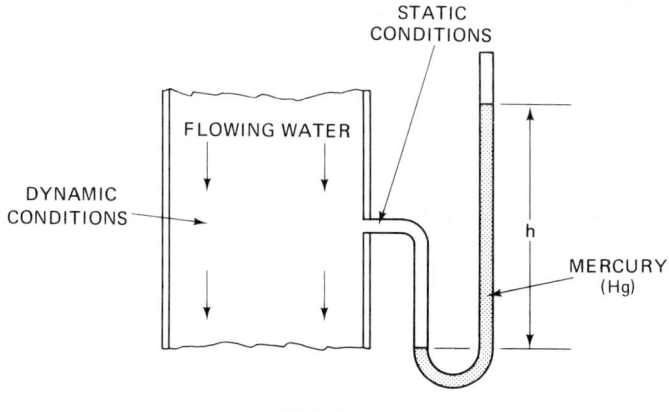

FIG. 6-3.

function in the same manner, since it is under static conditions in both cases. The pressure measurement is with respect to atmosphere and the manometer tube is open to the atmosphere at the top.

If the water pressure is removed from the pipe and the pressure at the manometer tap becomes equal to atmospheric pressure, the mercury assumes the position shown in Fig. 6-4. The height becomes zero and the pressure is atmospheric.

The use of manometers has given rise to alternative units of pressure measurement directly in inches or feet of the liquid used. These become equivalents to the more ordinary units of force per unit area, as exemplified by lb/in.^{2}.

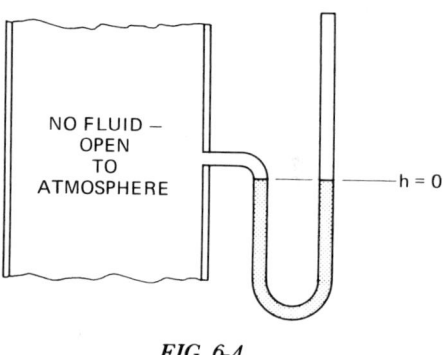

FIG. 6-4.

6-4 FLUID DENSITY AND PRESSURE

The *density* of a fluid has been defined as mass per unit volume. Mass, in turn, can be related to specific weight and specific gravity. Density is the fluid characteristic that directly affects the pressure at the base of a column of liquid. Since it is more convenient to work with specific weight rather than density, this term (W_s) will be used. In Fig. 6-5 a cylindrical glass tube with an inside cross-sectional area of 1 in.^{2} can be imagined to be filled at different times with four different liquids—water, mercury, alcohol, and lubricating oil. The depth of filling, h, is 4 in.

The table in Fig. 6-5 shows the calculated pressures at the bottom of the cylinder for each liquid:

$$P = W_s h \qquad (6\text{-}1)$$

where P, W_s, and h are as previously defined. Note that selection of consistent units in the W_s and h terms is important. To give pressure in

Sect. 6-4 / FLUID DENSITY AND PRESSURE

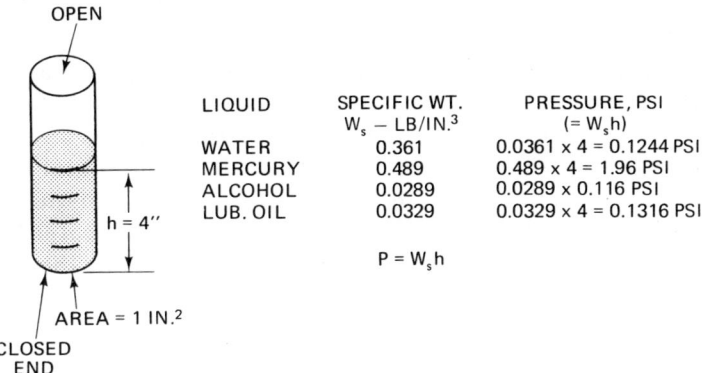

FIG. 6-5. Fluid weight—height—pressure relationship.

lb/in.^{2}, W_s must be in lb/in.^{3} and h in in. as follows:

$$\frac{\text{lb}}{\text{in.}^3}(\text{in.}) = \frac{\text{lb}}{\text{in.}^2}$$

If W_s and h are in feet, the pressure term will be lb/ft^2.

The constant of proportionality, k, discussed earlier now becomes the specific weight, W_s, of the liquid. Values of the specific weights of various liquids may be found in Table I in the Appendix.

EXAMPLE 1

The gravity tank in Fig. 6-6 supplies water to a distribution system for a manufacturing plant. Under static conditions what pressure would exist at a point on the third story of a building located within the plant? The elevations are as shown.

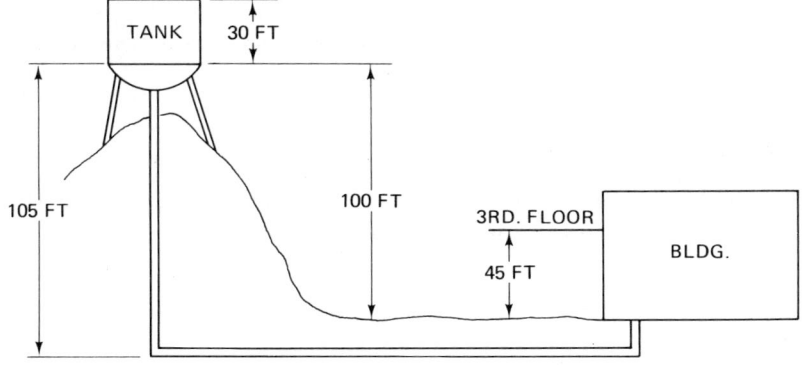

FIG. 6-6.

Solution

This essentially becomes a problem in analyzing elevations to determine the net head, h, which determines the pressure at the third-floor level. When this is done, Eq. (6-1), $P = W_s h$, applies.

By Pascal's law the pressure at all points in a horizontal plane is equal, and therefore the pressure at all points at the third-story level is the same. It is not known if the tank is full or if the water level is near the bottom. With the water level near the bottom, the head is computed as follows:

$$h = 100 \text{ ft} - 45 \text{ ft} = 55 \text{ ft}$$

Using Eq. (6-1), $P = W_s h$,

$$W_s = \frac{62.4 \text{ lb}}{\text{ft}^3}$$

Then

$$P = \frac{62.4 \text{ lb}}{\text{ft}^3} (55 \text{ ft}) = 3432 \frac{\text{lb}}{\text{ft}^2}$$

Note that P is in units of lb/ft². With the tank full the head h becomes $55 + 30 = 85$ ft. The corresponding pressure

$$P = (62.4)(85) = 5304 \frac{\text{lb}}{\text{ft}^2}$$

Note that the 105-ft distance from the tank to level of underground piping has no effect on the head at the third-story level.

EXAMPLE 2

Determine the pressure in lb/in.² at a point 35 ft below the surface of a lake.

Solution

Equation (6-1) applies and can be used to determine pressure at any point in a body of liquid open to the atmosphere.

$$P = W_s h \quad W_s = 62.4 \frac{\text{lb}}{\text{ft}^3}, \quad h = 35 \text{ ft}$$

$$= \left(62.4 \frac{\text{lb}}{\text{ft}^3}\right)(35 \text{ ft}) = 2184 \frac{\text{lb}}{\text{ft}^2}$$

In psi,

$$P = \frac{2134 \text{ lb/ft}^2}{144 \text{ in.}^2/\text{ft}^2} = 15 \frac{\text{lb}}{\text{in.}^2}$$

6-5 PRESSURE–HEIGHT RELATIONSHIP FOR GAS

In the preceding discussion, liquids were the only fluid considered. This was done because, in most cases, the specific weight of the gas involved is so small compared to the liquid that it can be neglected.

To illustrate this, consider the long vertical pipe shown in Fig. 6-7. The total length of pipe is 100 ft and the top 50 ft is filled with air under a pressure of 50 psig. The lower 50 ft is filled with water. The pressure at point A is 50 psig. The pressure at point B is 50 psig plus that due to the column of air above it, and the pressure at point C is 50 psig plus that due to the air and water above point C.

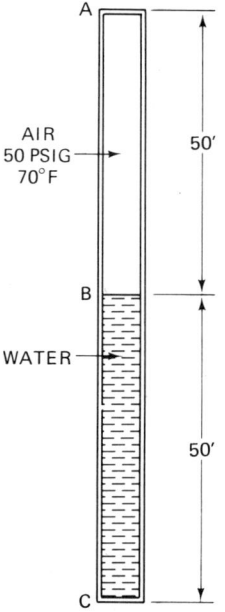

FIG. 6-7.

The air is assumed to have uniform density and the perfect-gas law is used to determine the specific volume, V_s. The reciprocal of the specific volume, $1/V_s$, is the specific weight. Then, using Eq. (4-3),

$$PV_s = RT$$

from Table IV, $R = 53.3 \dfrac{\text{ft lb}}{\text{lb °R}}$

$$P = (50 + 14.7)(144) = 9320 \dfrac{\text{lb}}{\text{ft}^2}$$

$$T = 70 + 460 = 530° \text{R}$$

$$V_s = \dfrac{RT}{P} = \dfrac{(53.3)(530)}{9320} = 3.03 \dfrac{\text{ft}^3}{\text{lb}}$$

But

$$\dfrac{1}{V_s} = W_s = \dfrac{1}{3.09} = 0.324 \dfrac{\text{lb}}{\text{ft}^3}$$

Using Eq. (6-1),

$P = W_s h$ at point B, $h = 50$ ft

$$= (0.324)(50) = 16.2 \dfrac{\text{lb}}{\text{ft}^2} = \dfrac{16.2}{144} \dfrac{\text{lb}}{\text{in.}^2}$$

$= 0.113$ psi at B due only to the weight of the column of air

At point C the pressure due to the weight of the column of the two fluids is

$$P = (\text{pressure of 50 ft of water}) + (\text{pressure of 50 ft of air})$$
$$= \dfrac{(62.4)(50)}{144} + 0.113 = 21.6 + 0.113 \text{ psi}$$

The total pressure at point C, P_{total}, is the sum of the 50 psi internal pressure and the pressures due to the column of fluids.

$$P_{\text{total}} = 50 + 21.6 + 0.113 = 71.7 \text{ psig}$$

Of this total, the column of air exerts only 0.1 psi pressure, a negligible quantity when compared to the total.

QUESTIONS AND PROBLEMS

6-1.—Calculate the pressure in lb/in.² at a depth of 175 ft in the ocean.

6-2.—A tank of oil with a specific gravity of 0.8 is 57 ft tall. What is the pressure at the bottom due to the oil when the tank is full?

6-3.—Compressed air maintains 30 psig pressure on the water in the storage tank in Fig. 6-8. The water level is kept constant at a

QUESTIONS AND PROBLEMS

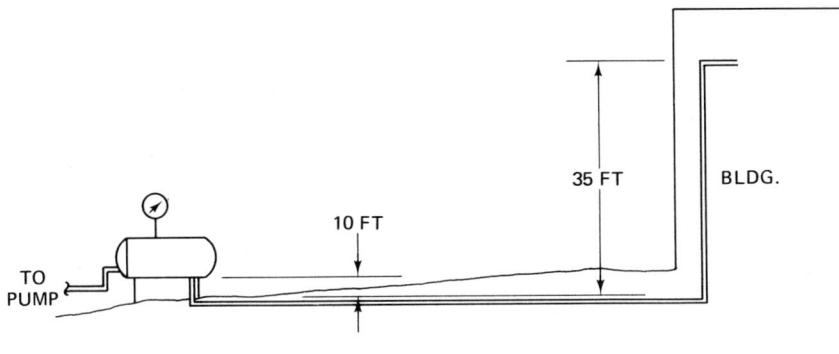

FIG. 6-8.

height of 2 ft above the tank bottom. What is the pressure at a point 35 ft above ground level in the building? Give your answer in psig and feet of water.

6-4.— Water and oil are stored in the tank in Fig. 6-9. The specific gravity of the oil is 0.83. Calculate the pressure on the tank bottom.

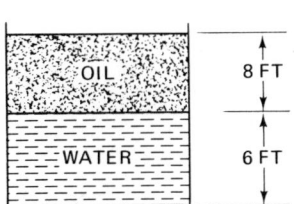

FIG. 6-9.

6-5.—The standpipe supplies water to the building in Fig. 6-10. The highest point in the building that water is required to flow is 40 ft above ground level. Is the standpipe adequate under all conditions to maintain pressure at this point? Support your answer by calculations.

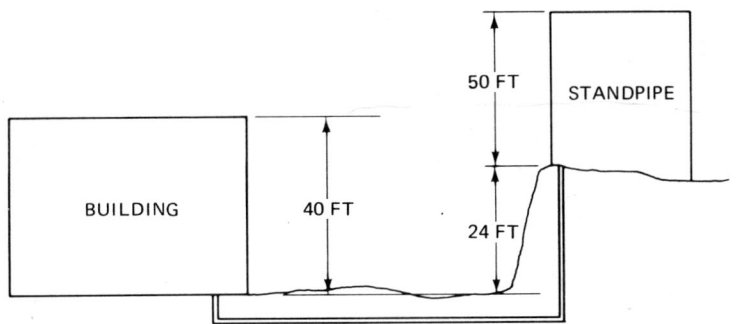

FIG. 6-10.

6-6.—Calculate the pressure at the bottom of a column of mercury 30 in. high.

6-7.— What is the pressure at the bottom of the container in Fig. 6-11? The fluid is water.

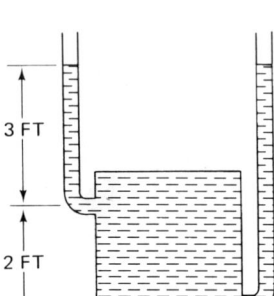

FIG. 6-11.

6-8.—Calculate the pressure at the bottom of a column of glycerine 26 in. high.

chapter seven
PRESSURE MEASUREMENT

7-1 INTRODUCTION

The measurement of pressure involves the application of the principles discussed in the preceding chapters. Any fluid system, whether it be the shop air system or the pressure-lubricating system of an internal combustion engine, generally has to have pressure-measuring devices, or gages, incorporated into it. In most cases the unit of measurement is psi, but there are other common units as well. These units are reviewed in the next section.

7-2 UNITS OF MEASUREMENT

Five common units of pressure measurement are listed in Table 7-1.

EXAMPLES:

$$1 \text{ psi} = 2.036 \text{ in. Hg}$$
$$1 \text{ atm} = 101.35 \text{ kPa}$$

Table 7-1   Pressure Equivalents

Pounds per Square Inch	Feet Water	Inches Mercury (in. Hg)	Atmospheres (atm)	Kilopascals (kPa)
1	2.309	2.036	0.068	6.89
0.433	1	0.882	0.0295	2.99
0.491	1.134	1	0.033	3.39
14.7	33.9	29.92	1	101.35
0.145	0.335	0.295	0.010	1

The kilopascal is used for the pressure unit in the SI system rather than the pascal (1 kPa = 1000 Pa) because of the small physical quantity represented by the Pa.

Pressure equivalents are in general interchangeable in calculations where pressures are involved as long as the other units in the calculations are consistent. There are other pressure equivalents in use in various specialized fields, but this text is restricted to those units in Table 7-1.

In measuring pressures below atmospheric, slightly different terminology is frequently used. In earlier discussions of absolute and gage pressures a diagram similar to Fig. 7-1 was used. In the vacuum region a negative gage pressure may be described as a vacuum. Inches of mercury is a frequently used unit of measurement, so a representative term such as 10 in. Hg vacuum means a negative gage pressure of 10 in. of mercury. If the pressure measurement were in psi, a pressure of 2 psi vacuum is a negative gage pressure of −2 psi. This method of specifying a vacuum is represented in Fig. 7-1.

Where very low pressures are involved, manometers measuring in inches of water may be used. In this case 1 in. of water is $\frac{1}{12}$ ft of water.

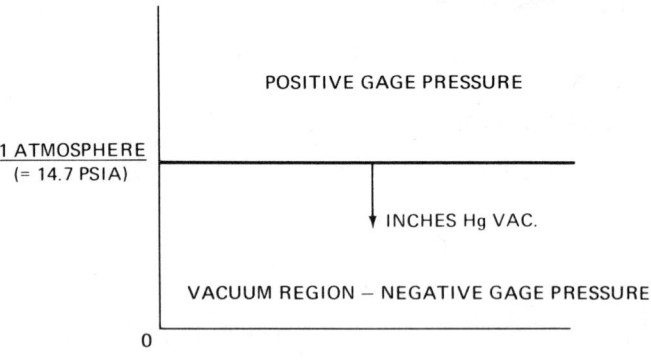

FIG. 7-1.

7-3 MANOMETERS

The *manometer* is a pressure-measuring device that makes use of the head or height of a column of liquid. In its common form it measures pressure with respect to atmosphere, either positive or negative. It can also be designed to read the difference in pressures between two fluids, both of which are under pressure. When used in this manner it is called a *differential manometer*. The mercury barometer uses the same principle to measure absolute pressure of the atmosphere.

Figure 7-2 shows a manometer attached to a pipe containing liquid 1 under a pressure P_1. The manometer also contains liquid 2, of specific gravity S_{g2}. The liquid in the pipe has a specific gravity of S_{g1}. The manometer is open to the atmosphere and therefore measures gage pressure. To analyze the system we proceed as follows:

1 / Since the manometer is a static system, all points within any one horizontal plane in the manometer have the same pressure. Therefore, points A and B have the same pressure. Note that point A was selected at the interface of the two liquids, to simplify the analysis.

2 / Write algebraic expressions for the pressure in the left-hand column and then the right-hand column. Equate these two expressions and solve for the desired pressure.

3 / This is illustrated below:

At point A, the pressure $= P_1 + W_{s1}h_1$ (W_{s1} is the specific weight of liquid 1)

At point B, the pressure $= W_{s2}h_2$ (W_{s2} is the specific weight of liquid 2)

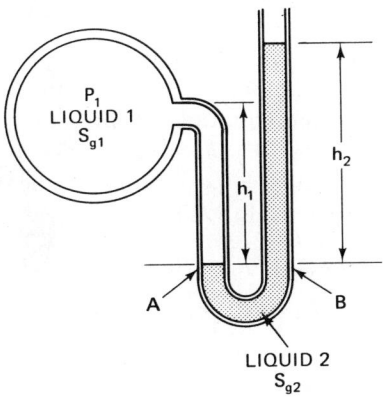

FIG. 7-2.

Equating the two,

$$P_1 + W_{s1}h_1 = W_{s2}h_2$$
$$P_1 = W_{s2}h_2 - W_{s1}h_1 \tag{7-1}$$

It is common practice to work with specific gravities of the liquids in a manometer. This can be done with Eq. (7-1) as follows. With W_s = specific weight of water, then $W_{s1} = S_{g1}W_s$ and $W_{s2} = S_{g2}W_s$. Substituting in Eq. (7-1),

$$P_1 = S_{g2}W_s h_2 - S_{g1}W_s h_1 = W_s(S_{g2}h_2 - S_{g1}h_1)$$
$$= W_s(S_{g2}h_2 - S_{g1}h_1) \tag{7-2}$$

The differential manometer is illustrated in Fig. 7-3. The manometer here is used to measure the drop in pressure at a restriction in the pipe through which a liquid is flowing. The liquid in the manometer is static, although the liquid in the pipe is under dynamic conditions.

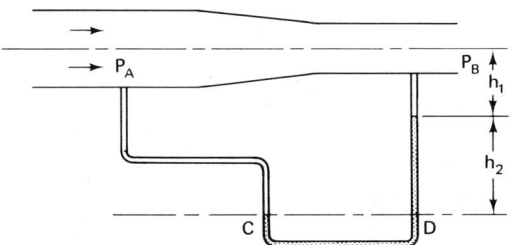

FIG. 7-3. Differential manometer.

The pressure difference that the manometer is required to measure is $P_A - P_B$. This pressure drop $(P_A - P_B)$ occurs because of the restriction in the pipe between A and B.

At points C and D in the manometer the pressures are equal since there is no flow and both points are in the same horizontal plane. An equation can then be written equating the pressures at these points. The pressure at point $C = P_A + W_{s1}(h_1 + h_2)$. The pressure at point $D = P_B + W_{s1}h_1 + W_{s2}h_2$. (Subscripts 1 and 2 indicate the liquids in the pipe and manometer, respectively.) Then,

$$P_A + W_{s1}(h_1 + h_2) = P_B + W_{s1}h_1 + W_{s2}h_2$$
$$P_A - P_B = W_{s1}h_1 + W_{s2}h_2 - W_{s1}(h_1 + h_2)$$
$$= W_{s1}h_1 + W_{s2}h_2 - W_{s1}h_1 - W_{s1}h_2$$
$$= W_{s2}h_2 - W_{s1}h_2$$
$$= h_2(W_{s2} - W_{s1}) \tag{7-3}$$

Sect. 7-3 / MANOMETERS

Equation (7-3) can be converted to use specific gravities and the specific weight of water. Using W_s for the specific weight of water,

$$P_A - P_B = h_2(S_{g2}W_s - S_{g1}W_s) = W_s h_2(S_{g2} - S_{g1})$$
$$= W_s h_2(S_{g2} - S_{g1}) \qquad (7\text{-}4)$$

Care should be taken in using these equations, since slight differences in the way the manometer is used may cause the equation to be invalid. If the pipe in Fig. 7-3 were not horizontal and point B were higher than point A, the difference in elevation would have to be taken into consideration. If the fluid being measured by the manometer is a gas, then the terms relating to the specific gravity and specific weight drop out of the equations, leaving only the manometer liquid terms in the equations.

Equations (7-1) through (7-4) are applicable to some but not all of the same problems that follow. Consequently, each problem should be analyzed carefully before proceeding with it.

EXAMPLE 1

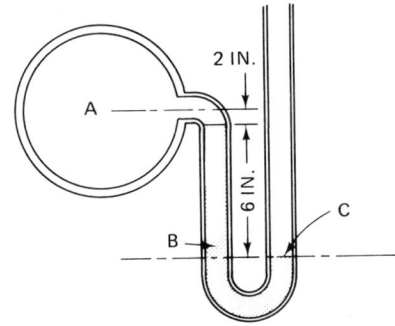

FIG. 7-4.

The pipeline shown in Fig. 7-4 has within it a liquid of 0.8 specific gravity. The manometer shown uses mercury with a specific gravity of 13.6. With the conditions as shown, determine the pressure at point A.

Analysis

The manometer is similar to that shown in Fig. 7-2 except that the long column of mercury is on the left instead of the right. Instead of trying to use Eq. (7-2), which applies for Fig. 7-2, a new one will be developed. The pressures at points B and C are equal, but P_c is atmospheric pressure. Since the equation is written in terms of gage pressure, then $P_c = 0$. The specific weight of the liquid is equal to 0.8 W_s, where W_s is the specific weight of water. In terms of the column on the left-hand side the pressure at B is:

$$P_B = P_A + (2)(0.8)W_s + (6)(13.6)W_s$$

But $P_B = P_c = 0$; then

$$P_A + (2)(0.8)W_s + (6)(13.6)W_s = 0$$
$$P_A = (-2)(0.81)W_s - (6)(13.6)W_s$$

and with $W_s = 0.0361$ lb/in.^{3},

$$P_A = (-2)(0.8)(0.0361) - (6)(13.6)(0.0361)$$
$$= -0.0577 - 2.95 = -3.01 \text{ psig}$$

The pressure at point A is actually a vacuum.

EXAMPLE 2

The tank in Fig. 7-5 contains water under a pressure that causes it to rise to a height of 8 in. above point A. What is the pressure at A?

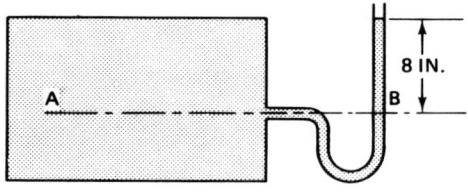

FIG. 7-5.

Analysis

Equation (7-1) can be used to solve the problem. The equation is

$$P_A = W_{s2}h_2 - W_{s1}h_1$$

If points A and B are selected as reference, h_1 becomes zero (see Fig. 7-2) and $W_{s1} = W_{s2}$ = specific weight of water. Then

$$P_A = W_s h_2 = (0.0361)(8)$$
$$= 0.29 \text{ psig}$$

Another acceptable answer to this is simply 8 in. of water, since inches and feet of water are alternative units of pressure measurement. The answer then can be obtained by inspection.

EXAMPLE 3

The differential manometer in Fig. 7-6 measures the pressure difference between A and B. Water flows in the pipe in the direction indicated and the manometer fluid is mercury. Determine $P_A - P_B$.

Sect. 7-4 / BOURDON-TUBE GAGE

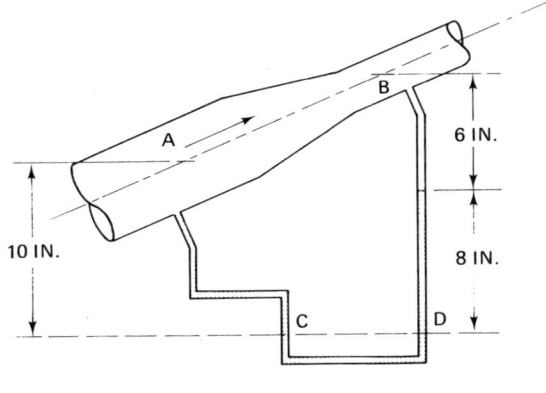

FIG. 7-6.

Analysis

The system is complicated enough so that its own equation should be written. Points C and D are selected in the same horizontal plane. At point C the pressure is $P_A + 10$ in. water; at point D the pressure is $P_B + 8$ in. Hg + 6 in. water. Equating these to each other and converting to common terms,

$$P_A + 10W_s = P_B + (8)(13.6)W_s + 6W_s$$

where W_s = specific weight of water = 0.0361 lb/in.^{3}. Then

$$P_A - P_B = (8)(13.6)(0.0361) + (6)(0.0361) - (10)(0.0361)$$
$$= 3.8 \text{ psig}$$

7-4 BOURDON-TUBE GAGE

The most common type of pressure gage in use is the Bourdon-tube pressure gage. This type of gage is mechanical in operation and depends on the deflection of one end of a circular tube when pressure is applied. Its principle of operation is shown schematically in Fig. 7-7. The flattened tube is the pressure chamber, which tends to straighten out when pressure is applied. The movement moves the link and causes the gear sector, pivoted at A, to rotate. The remainder of the movement is through gearing and a spiral spring to the pointer, which indicates the pressure on the face of the dial.

The gage measures with reference to atmosphere, since there is no deflection of the tube until a pressure different from atmosphere is applied. Since negative pressure will cause the tube to deflect in the opposite direc-

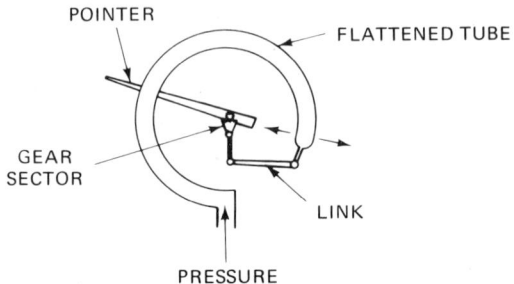

FIG. 7-7. Mechanism of the Bourdon-tube gage.

tion, it can be used for measuring vacuum as well as positive pressure if designed for this purpose.

By suitable design of the tube element and gearing, Bourdon gages are made in varying pressure ranges, from 0–10 psi to 0–1000 psi and higher ranges.

7-5 *STATIC PRESSURE MEASUREMENT UNDER DYNAMIC CONDITIONS*

Both manometers and Bourdon-tube pressure gages are used to measure static pressure in fluid systems that are under dynamic (flow) conditions. It is possible to measure static pressure in a dynamic condition if the gage

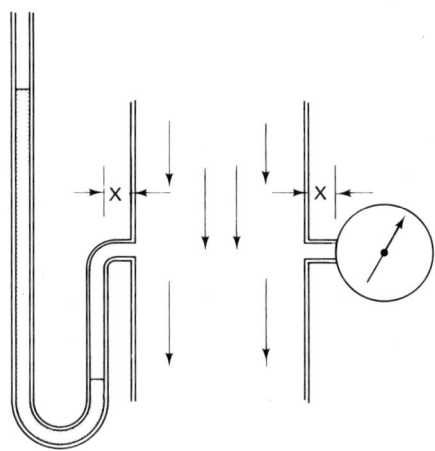

FIG. 7-8. Measurement of static pressure under flow conditions.

is installed in such a manner that static conditions exist in the gage and in immediate surroundings to the gage. In Fig. 7-8, fluid is flowing vertically downward in a pipe. A manometer and a Bourdon-tube pressure gage are installed on opposite sides of the pipe to measure static pressure.

In order to assure static conditions in the gages the pressure taps at A and B are at right angles to the direction of flow. The hole for the tap should be small, and increasing the distance X will provide additional assurance of static conditions.

QUESTIONS AND PROBLEMS

7-1.—Convert 12.5 psi to the following.
  a—Atmospheres.
  b—Inches of mercury.
  c—Feet of water.
  d—Kilopascals.

7-2.—The pressure in a system is indicated by the height of a column of liquid of 1.26 specific gravity. The height measures 2.6 ft. What is the pressure in the following units?
  a—psi.
  b—ft of water.
  c—in. Hg.

7-3.—The tank in Fig. 7-9 is filled with water and the manometer liquid is mercury. Determine the pressure at the top of the tank.

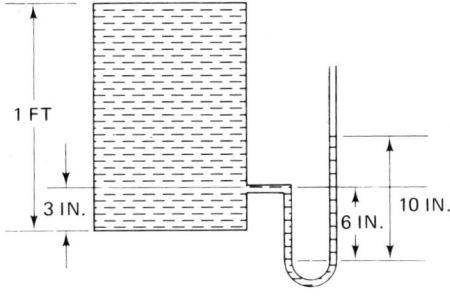

FIG. 7-9.

7-4.—Convert a pressure of 12 in. Hg vacuum to psi absolute.

7-5.—Convert 6 in. Hg vacuum to gage pressure in psi.

7-6.—

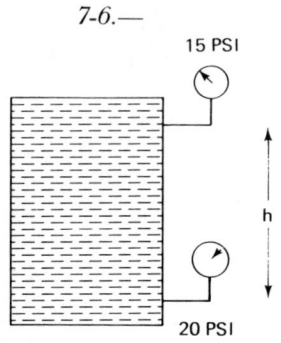

FIG. 7-10.

The long vertical storage tank in Fig. 7-10 contains water under static conditions. The top pressure gage reads 15 psi and the lower one 20 psi. What is the distance h?

7-7.—In the same storage tank of Problem 7-6, what is the distance h if the liquid is oil with a specific gravity of 0.86?

7-8.—

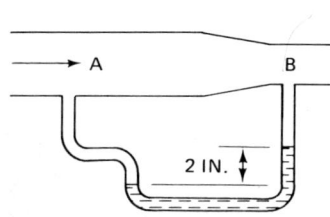

FIG. 7-11.

Water flows in the pipe in Fig. 7-11 with the restriction as shown. The manometer contains mercury. Determine the difference in pressure between points A and B.

7-9.—

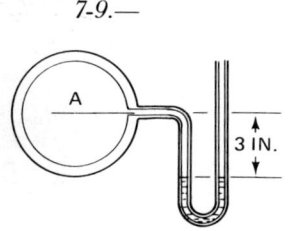

FIG. 7-12.

What is the pressure at point A in the static system in Fig. 7-12? The fluid in the pipe is water and that in the manometer is mercury.

7-10.—Water flows in the pipe as shown in Fig. 7-13 and the manometer fluid is mercury. Determine the pressure difference between A and B if B is 3 ft higher in elevation than A.

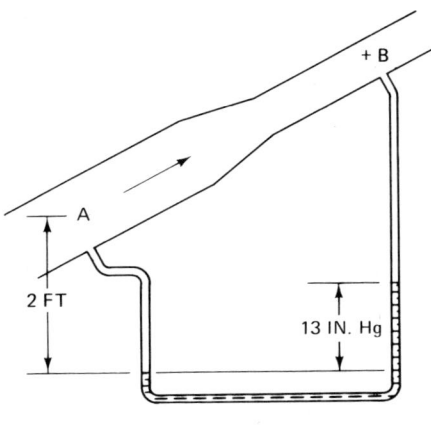

FIG. 7-13.

chapter eight

FLUID DYNAMICS— INCOMPRESSIBLE FLUIDS

8-1   INTRODUCTION

In this chapter fluid dynamics is considered primarily from a theoretical standpoint. In later chapters this theoretical approach is modified so as to make it usable for practical problems involving fluid dynamics and fluid flow. The theory of fluid flow rests on two main principles. These are:

1 /   **The principle of conservation of mass.**
2 /   **The principle of conservation of energy.**

Briefly described, these two principles state that mass and energy may exist in different forms in a system but that the total quantity of either mass or energy will remain constant in the system.

The conditions which this discussion will apply to are those which are called *steady-state flow conditions*. Steady-state flow occurs when conditions do not change with time. Flow through a pipe with a valve fully open is steady state when, at any given location, the pressure, flow rate, temperature, and any other variable affecting the process remain constant. These variables may have different values at other locations, but at each location they are constant. If the valve opening is changed, while it is being changed the flow is unsteady, since the variables are changing with respect to time.

8-2 PRINCIPLE OF CONSERVATION OF MASS

The analysis of steady-state flow is made with the use of assumed streamlines of flow, as shown in Fig. 8-1. The streamlines of flow are uniform and converge or diverge as the tube converges or diverges.

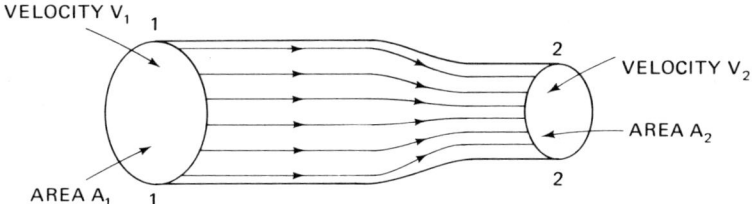

FIG. 8-1. Streamline flow.

Using the principle of conservation of mass it can be said that the same mass of fluid that flows past section 1-1 in Fig. 8-1 in a given period of time will also flow past section 2-2 in the same period of time. This becomes the basis for setting up the mathematical equations governing flow velocities and quantities for a fluid. If A_1 equals the cross-sectional area at section 1-1 and v_1 equals the average velocity of the fluid passing the section, the volume passing the section is $A_1 v_1$. If units of feet and seconds are used, this becomes $(\text{ft}^2)(\text{ft/s}) = \text{ft}^3/\text{s}$. If the quantity $A_1 v_1$ is multiplied by the density ρ, the result is the mass rate of flow.

Since it is easier to work in weight units, the specific weight W_s can be used instead of ρ. When this is multiplied by $A_1 v_1$, the result is $W_s A_1 v_1$. The units of this are $(\text{lb/ft}^3)(\text{ft}^2)(\text{ft/s})$. These reduce to lb/s.

Since no fluid is being lost or added to the tube between sections 1-1 and 2-2, the same weight or mass must flow past section 2-2. Therefore,

$$W_s A_1 v_1 = W_s A_2 v_2 \tag{8-1}$$

where A_2 and v_2 are the area and velocity at section 2-2.

Equation (8-1) is termed the *equation of continuity in terms of weight*. For incompressible fluids (liquids) where the specific weight W_s does not change, the term W_s can be cancelled from Eq. (8-1), leaving $A_1 v_1 = A_2 v_2$. If the product of the area and velocity is called Q, the volume flow

Sect. 8-2 / PRINCIPLE OF CONSERVATION OF MASS

rate, the equation may be rearranged as follows:

$$Q = A_1 v_1 = A_2 v_2 = A_3 v_3 = \cdots \qquad (8\text{-}2)$$

Equation (8-2) is the *equation of continuity in terms of volume rate of flow*.

EXAMPLE 1

Ten gallons per minute of water is flowing in a pipe with an inside diameter of 2 in. What is the average velocity of the water in the pipe?

Solution

From Table V in the Appendix, 1 gal = $1/7.48$ ft^3 and 10 gal = $10/7.48 = 1.34$ ft^3. Using Eq. (8-2), $Q = Av$ and $v = Q/A$.

$$A = \frac{\pi}{4} \left(\frac{2}{12}\right)^2 = \frac{\pi}{4} \frac{1}{36} = 0.0218 \text{ ft}^2$$

Note the need to change the area to units consistent with that for the velocity. Then,

$$v = \frac{1.34 \text{ ft}^3/\text{min}}{0.0218 \text{ ft}^2} = 61.5 \frac{\text{ft}}{\text{min}}$$

EXAMPLE 2

A quantity of water is flowing in the pipe in Fig. 8-2. What is the ratio of the velocity at point 2 to that at point 1?

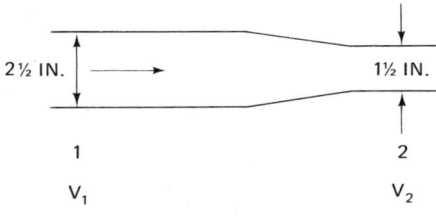

FIG. 8-2.

Solution

The ratio desired is v_2/v_1. This can be obtained by using Eq. (8-2) again and arranging terms as follows:

$$A_1 v_1 = A_2 v_2 \quad \text{or} \quad v_2 = \frac{A_1 v_1}{A_2}$$

8-3 PRINCIPLE OF THE CONSERVATION OF ENERGY

The principle of the conservation of energy states that the total energy in any system remains constant. The explanation of this as related to fluid mechanics requires the use of Bernoulli's equation plus the use of two new energy terms, potential energy and kinetic energy.

Potential energy may be considered as the energy a mass contains by virtue of its position. Kinetic energy is the energy a mass contains by virtue of its motion or velocity. Figure 8-3 illustrates these two energy types. In Fig. 8-3(a) is shown a hypothetical 1 lb of fluid, which is a distance of Z ft above a datum plane. In Fig. 8-3(b) is shown a hypothetical 1 lb of fluid, which is moving with a velocity v in a pipe. Both of these energy forms may be put into equations, as follows:

potential energy = (weight of object)(distance above plane)

kinetic energy = $\frac{1}{2}$(mass)(velocity)^2

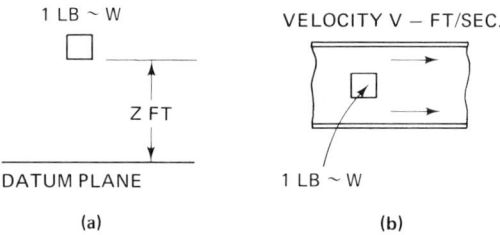

FIG. 8-3. Potential and kinetic energy.

Using PE and KE to represent potential and kinetic energy, respectively, and the symbols from Fig. 8-3, the equations become

$$PE = WZ \qquad (8\text{-}3)$$

$$KE = \tfrac{1}{2} Mv^2 \qquad (8\text{-}4)$$

In 1738 Daniel Bernoulli used these terms to help develop a basic energy equation called simply "Bernoulli's equation," which is still used today. Before getting into the equation itself, it is well to consider the units of energy that will be used. These units of energy can be developed from Eqs. (8-3) and (8-4). In Eq. (8-3), W represents 1 lb of incompressible fluid and Z the elevation or height in feet. The product of these is ft lb of energy. Since this has been restricted to 1 lb of fluid, the statement can be resolved mathematically as

$$\text{ft lb per lb} \quad \text{or} \quad \frac{\text{ft lb}}{\text{lb}}$$

Sect. 8-3 / PRINCIPLE OF THE CONSERVATION OF ENERGY

The kinetic-energy term can be developed in a similar manner. To do this requires the substitution of W/g for the mass M. Then $M = W/g$ and $KE = \frac{1}{2}(W/g)v^2$. Using the regular units of feet, lb, and velocity in ft/s (g is in units of ft/s^2),

$$KE = \frac{\text{lb}}{\text{ft/s}^2}\left(\frac{\text{ft}}{\text{s}}\right)^2 = \frac{\text{lb ft}^2\text{s}^2}{\text{ft s}^2} = \text{ft lb}$$

Since the term above applies to 1 lb of fluid, it can be restated as ft lb/lb. The units are now the same as the units for potential energy.

Now consider the system shown in Fig. 8-4. Here there is an incompressible fluid flowing in a pipe with different elevations and pipe diameters at points 1 and 2. The average velocities, v_1 and v_2, are different because of the different diameters. The fluid is also under a pressure P_1 at point 1 and P_2 at point 2. Since the internal pressure of any fluid system is capable of doing work, it also represents an energy source. It is another potential-energy form, since by definition the kinetic-energy forms are all represented in the fluid velocity.

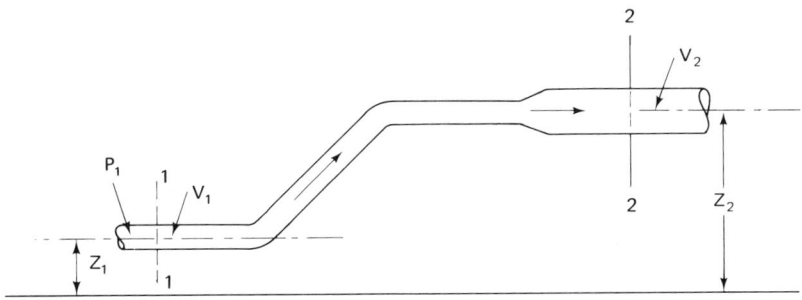

FIG. 8-4.

This internal pressure term can be converted to the same energy units as velocity and elevation by considering the work done by pressure acting on a volume of fluid as shown in Fig. 8-5. Here pressure provides a force that moves an increment of volume of a fluid through a distance d in a pipe. This increment of volume in turn has forced the preceding fluid volume into the next pipe segment and consequently has done work on it. The work done by the fluid moving a distance d with a force F acting on it $= Fd$. Since force = pressure $\times$ area, then $F = PA$ and the work $= PAd$. The weight of fluid that does this work is the specific weight $\times$ volume, or W_sAd. Then, the work or energy per pound of fluid is

$$\frac{PAd}{W_sAd} = \frac{P}{W_s}$$

Checking units, P in lb/ft² and W_s in lb/ft³,

$$\frac{\text{lb/ft}^2}{\text{lb/ft}^3} = \frac{\text{ft lb}}{\text{lb}}$$

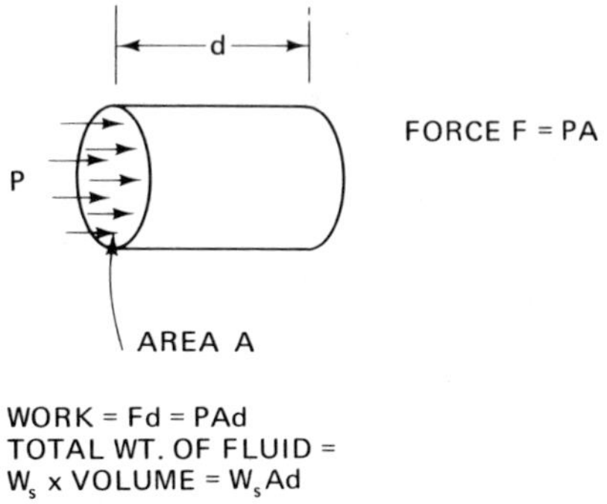

FIG. 8-5.

Referring again to Fig. 8-4, there are now three energy forms at each of the positions 1 and 2. Since the total energy at points 1 and 2 is the same (no energy is added or taken away between these points), an equation can be written for the conditions. The equation will be written for 1 lb of fluid. The kinetic-energy term will then reduce to $\frac{1}{2}(v^2/g)$ or $v^2/2g$ when it is divided by W, the total weight. The equation then is

$$\frac{P_1}{W_s} + \frac{v_1^2}{2g} + Z_1 = \frac{P_2}{W_s} + \frac{v_2^2}{2g} + Z_2 \qquad (8\text{-}5)$$

Equation (8-5) is Bernoulli's equation in its most common form. Note that the units of each term are ft lb/lb of fluid flowing. This is consistent with the energy definitions that have been used. When the term is written in fractional form, ft lb/lb, it will be noted that the weight unit, lb, will cancel out, leaving only the units of elevation, ft. In practice, then, Bernoulli's equation can be thought of in energy units and also pressure units. This is true because pressure measurements are made in elevation (or heights) as

Table 8-1

Mathematical Term	Name
$\dfrac{P}{W_s}$	Pressure head
$\dfrac{v^2}{2g}$	Velocity head
Z	Potential or elevation head

well as force per unit area. Thus, an elevation term, Z, in height can be used directly in the equation. Bernoulli's equation is so important and used so much in fluid mechanics that each of the three energy terms has been given its own descriptive name. They are listed in Table 8-1.

8-4 ENERGY ADDITIONS AND LOSSES

The form of Bernoulli's equation developed here assumes that no energy is gained or lost in the system. This is a theoretical approach that has to be modified to fit actual conditions. For instance, a pump can be put into a system to boost the pressure head and this is an example of energy added to the system. In practice, all flowing systems have energy losses, which are termed *friction losses*. Revising Bernoulli's equation, it can be stated as follows:

$$\begin{bmatrix}\text{press.}\\\text{head}\end{bmatrix}_1 + \begin{bmatrix}\text{vel.}\\\text{head}\end{bmatrix}_1 + \begin{bmatrix}\text{elev.}\\\text{head}\end{bmatrix}_1 + \begin{bmatrix}\text{energy}\\\text{added}\end{bmatrix}_{1\text{-}2}$$
$$= \begin{bmatrix}\text{press.}\\\text{head}\end{bmatrix}_2 + \begin{bmatrix}\text{vel.}\\\text{head}\end{bmatrix}_2 + \begin{bmatrix}\text{elev.}\\\text{head}\end{bmatrix}_2 + \begin{bmatrix}\text{energy}\\\text{lost}\end{bmatrix}_{1\text{-}2}$$

Using Fig. 8-4 again, the energy added is at some location between points 1 and 2. The mathematical terms corresponding to these descriptive terms is developed in later units.

EXAMPLE 1

The reservoir in Fig. 8-6 supplies water with a constant head h of 42 ft. The pipe is 2 in. and 4 in. inside diameter for the distances shown. Write Bernoulli's equation for (a) points 1 and 2 and (b) points 1 and 3. Use the datum plane as shown.

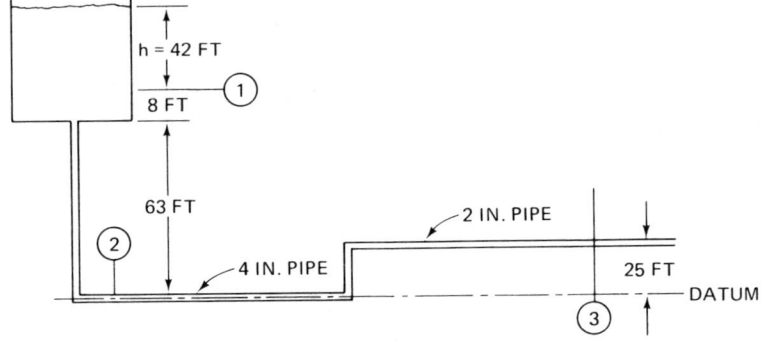

FIG. 8-6.

Solution

Taking points 1 and 2 first, at point 1 the pressure head P_1/W_s is that caused by the constant head h of 42 ft. This is already in feet, so it can be used as is in the equation. At point 1, the velocity term is $v_1^2/2g$. The elevation term at point 1 is the elevation at point 1, or $8 + 63 = 71$ ft. At point 2 the elevation head is zero and the pressure and velocity heads are unknown. Using Eq. (8-5) and substituting the values as noted gives

$$42 + \frac{v_1^2}{2g} + 71 = \frac{P_2}{W_s} + \frac{v_2^2}{2g} + 0 \qquad \text{points 1 and 2}$$

At point 3 the elevation head is 25 ft and the velocity head and pressure head are unknown. The equation for points 1 and 3 then is

$$42 + \frac{v_1^2}{2g} + 71 = \frac{P_3}{W_s} + \frac{v_3^2}{2g} + 25 \qquad \text{points 1 and 3}$$

The problem asks for the equation for the points, and these are provided. Note that no attempt has been made to simplify them, although some of the terms could have been reduced. If the volume rate of flow were known, the equation of continuity [Eq. (8-2)] could be used to determine the velocities at points 2 and 3, since the pipe diameters are known at these points. At point 1 the area of the reservoir is not known, but it is reasonable to assume that it is large in comparison with the pipe cross-sectional area. Under these conditions, then, the velocity at point 1 is so small that it can be neglected.

EXAMPLE 2

Twenty gallons per minute of oil with a specific gravity of 0.8 flows in the horizontal pipeline in Fig. 8-7. The pressure gage at A reads 35 psi. What does the gage at B read?

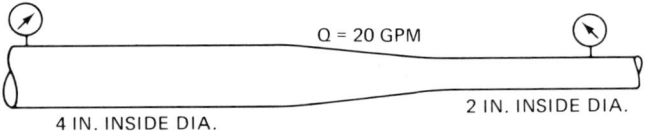

FIG. 8-7.

Solution

Bernoulli's equation is used and the two locations are designated A and B. Since the pipe is horizontal, the elevation of both locations from any reference plane is the same and the terms can be dropped

Sect. 8-4 / ENERGY ADDITIONS AND LOSSES

from the equation. The equation then becomes

$$\frac{P_A}{W_s} + \frac{v_A^2}{2g} = \frac{P_B}{W_s} + \frac{v_B^2}{2g}$$

Rearranging,

$$\frac{P_B}{W_s} = \frac{P_A}{W_s} + \frac{v_A^2}{2g} - \frac{v_B^2}{2g}$$

$$P_B = W_s\left(\frac{P_A}{W_s}\right) + W_s\left(\frac{v_A^2}{2g} - \frac{v_B^2}{2g}\right)$$

$$= P_A + \frac{W_s}{2g}(v_A^2 - v_B^2)$$

W_s, v_A, and v_B are needed and are determined as shown.

$$W_s = 0.8\left(62.4 \frac{\text{lb}}{\text{ft}^3}\right) = 49.9 \frac{\text{lb}}{\text{ft}^3}$$

From Eq. (8-2), the velocity $v = Q/A$ and

$$Q = \frac{20 \text{ gal/min}}{7.48 \text{ gal/ft}^3} = 2.67 \frac{\text{ft}^3}{\text{min}}$$

The area of the 4-in. pipe is $\pi/4\,(4/12)^2 \text{ ft}^2 = 0.087 \text{ ft}^2$

$$v_A = \frac{2.67 \text{ ft}^3/\text{min}}{0.087 \text{ ft}^2} = 30.7 \frac{\text{ft}}{\text{min}} = 0.52 \frac{\text{ft}}{\text{s}}$$

The area of the 2-in. pipe is $(\pi/4)(2/12)^2 \text{ ft}^2 = 0.0218 \text{ ft}^2$

$$v_B = \frac{2.67 \text{ ft}^3/\text{min}}{0.0218 \text{ ft}^2} = 122.5 \frac{\text{ft}}{\text{min}} = 2.04 \frac{\text{ft}}{\text{s}}$$

The values may now be used in the equation that has been determined for P_B. With g in ft/s^2 and P_A in lb/ft^2, all units are consistent.

$$P_B = \left(35 \frac{\text{lb}}{\text{in.}^2}\right)\left(144 \frac{\text{in.}^2}{\text{ft}^2}\right) + \frac{49.9 \text{ lb/ft}^3}{2\,(32.2 \text{ ft/s}^2)}\left[\left(0.52 \frac{\text{ft}}{\text{s}}\right)^2 - \left(2.04 \frac{\text{ft}}{\text{s}}\right)^2\right]$$

$$= 5040 \frac{\text{lb}}{\text{ft}^2} + \frac{49.9}{64.4}(-3.89) \frac{\text{lb}}{\text{ft}^2}$$

$$= 5037 \frac{\text{lb}}{\text{ft}^2} = 34.9 \frac{\text{lb}}{\text{in.}^2}$$

This change from the pressure at A is so small that it probably would not be indicated in the gage at B. The problem does, however, illustrate that a change in velocity head changes the pressure head.

8-5 TORRICELLI'S THEOREM

The theoretical velocity of a free jet of fluid discharging into the atmosphere can be determined by the application of Bernoulli's equation. Consider the reservoir shown in Fig. 8-8. A free jet discharges from the circular opening into the atmosphere. Bernoulli's theorem is applied at points 1 and 2, and the head h is constant. With a large surface area at point 1, the velocity at this point can be considered to be zero. Since atmospheric pressure is being used as the reference, the pressure head at both points is zero. Bernoulli's equation is

$$\frac{P_1}{W_s} + \frac{v_1^2}{2g} + Z_1 = \frac{P_2}{W_s} + \frac{v_2^2}{2g} + Z_2$$

When the values are substituted in this, there is

$$0 + 0 + h = 0 + \frac{v_2^2}{2g} + 0$$

Solving for v_2, then

$$v_2 = \sqrt{2gh} \tag{8-6}$$

This mathematical statement is known as *Torricelli's theorem*.

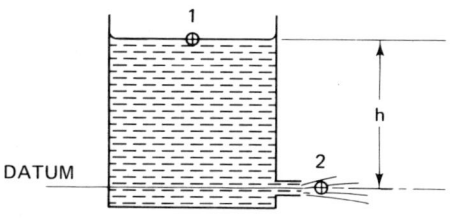

FIG. 8-8.

QUESTIONS AND PROBLEMS

8-1.—Twenty-five gallons of water per minute is flowing in a pipeline. What is the flow rate in cubic feet per second?

8-2.—Fifty-three cubic feet per second of water flows in the pipe in Fig. 8-9. With the diameters as shown, determine the velocities at points 1, 2, and 3.

QUESTIONS AND PROBLEMS

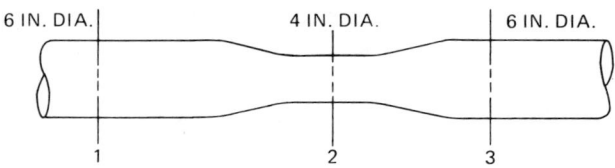

FIG. 8-9.

8-3.—In Problem 8-2, how many lb/s is flowing?

8-4.—

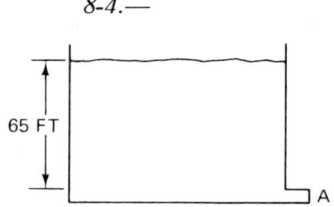

The reservoir in Fig. 8-10 is kept with a constant 65-ft head. The fluid is water. Determine the velocity of the water that flows through the opening at point A.

FIG. 8-10.

8-5.—The pipe in Fig. 8-11 has 250 gpm (gal per min) of water flowing through it. Determine the pressure at point B. (*Note:* Pressure described in this manner is the pressure-head term in Bernoulli's equation.)

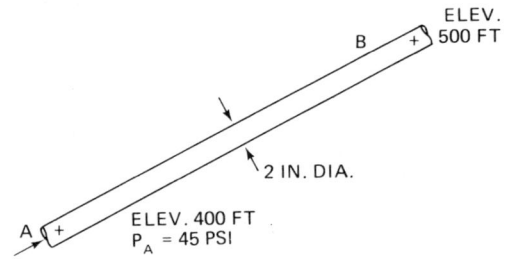

FIG. 8-11.

8-6.—

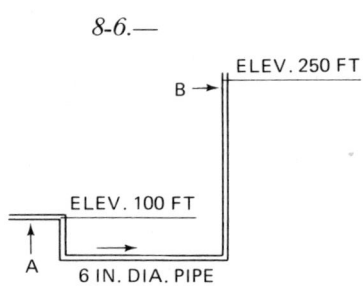

With $Q = 156$ ft^3/s and oil with specific gravity of 0.86, determine the pressure difference between points A and B in Fig. 8-12. (*Note:* The pressure difference is $P_A - P_B$.)

FIG. 8-12.

8-7.—What pipe diameter is required to carry 36 gpm of water with a velocity of 10 ft/s?

8-8.—This closed tank is pressurized at 15 psig by air as shown in Fig. 8-13. The liquid is water. Determine the velocity of the water issuing from the opening at the bottom.

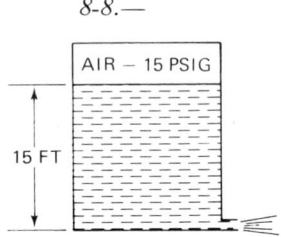

FIG. 8-13.

8-9.—The volume flow rate of a certain fluid in a pipeline is 255 gpm. If the specific gravity of the fluid is 0.92, what is the weight flow rate in lb/s?

8-10.—A vacuum pump maintains a vacuum of −3 ft of water at point A in the reservoir in Fig. 8-14. With the tank left open to the atmosphere, determine the flow rate of the water through the 2-in.-diameter pipe. What is the direction of flow?

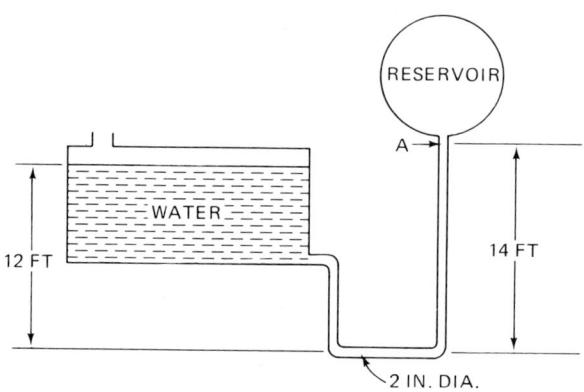

FIG. 8-14.

chapter nine

FLOW TYPES AND FRICTION LOSS— INCOMPRESSIBLE FLUIDS

9-1   INTRODUCTION

In Chapter 8 it was stated that the practical use of Bernoulli's equation requires consideration of any energy added to the system plus energy losses, which are called *friction losses*. Restating the equation, it becomes

$$\begin{bmatrix}\text{press.}\\\text{head}\end{bmatrix}_1 + \begin{bmatrix}\text{vel.}\\\text{head}\end{bmatrix}_1 + \begin{bmatrix}\text{elev.}\\\text{head}\end{bmatrix}_1 + \begin{bmatrix}\text{energy}\\\text{added}\end{bmatrix}_{1-2}$$

$$= \begin{bmatrix}\text{press.}\\\text{head}\end{bmatrix}_2 + \begin{bmatrix}\text{vel.}\\\text{head}\end{bmatrix}_2 + \begin{bmatrix}\text{elev.}\\\text{head}\end{bmatrix}_2 + \begin{bmatrix}\text{energy}\\\text{loss}\end{bmatrix}_{1-2} \quad (9\text{-}1)$$

This principle is illustrated in Fig. 9-1. Here a pump adds energy to the system by pumping a fluid from an open reservoir through a piping system, which discharges back into the atmosphere. Between point 1 and the pump, there is an energy loss due to pipe friction. Between the pump and point 2 there is another loss due to friction. The energy added to the system is at the pump itself.

Now Eq. (9-1) can be rewritten with the mathematical symbols as

82 FLOW TYPES AND FRICTION LOSS—INCOMPRESSIBLE FLUIDS / Chap. 9

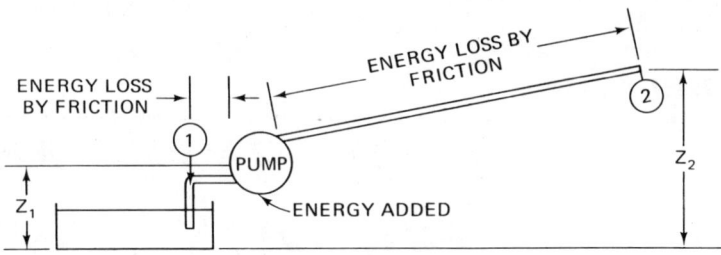

FIG. 9-1.

used in Chapter 8, and symbols can be added for the other two energy terms. It then becomes

$$\frac{P_1}{W_s} + \frac{v_1^2}{2g} + Z_1 + h_p = \frac{P_2}{W_s} + \frac{v_2^2}{2g} + Z_2 + h_L \qquad (9\text{-}2)$$

In order to use Eq. (9-2) the units of h must be consistent with the other terms in the equation. In energy units the other terms are in ft lb/lb of fluid flowing. Therefore, h_L must be able to be determined in the same units. This can be done and the methods for determining the friction loss, h_L, are developed in this chapter. Later chapters cover pumps in more detail.

9-2 LAMINAR FLOW

Osborne Reynolds, an English scientist, classified the flow of fluids into two categories, which are dependent upon the velocity of the fluid. The terms used to describe the two flow conditions are laminar and turbulent. Laminar flow may be described as viscous, streamline, or nonsinuous motion of the fluid particles. It can best be visualized by referring to Fig. 9-2. In laminar flow conditions a stream of dye injected into the flow of water in the pipe does not mix with the water, and the stream continues in a straight line, as shown.

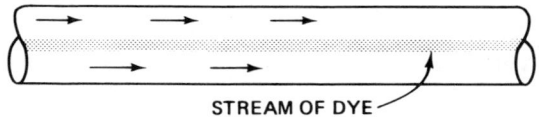

FIG. 9-2. Laminar flow.

Now, if the velocity of the fluid in the pipe is increased to a certain value, the streamline of dye in Fig. 9-2 begins to become sinuous, wavy, and then break up, so the dye particles mix with the fluid particles. This becomes the onset of turbulent flow.

9-3 TURBULENT FLOW

Figure 9-3 shows a representation of turbulent-flow conditions with the same stream of dye. The stream of dye first begins to become wavy and then begins to break up, until it finally mixes with the primary fluid in the system. This is *turbulent flow*.

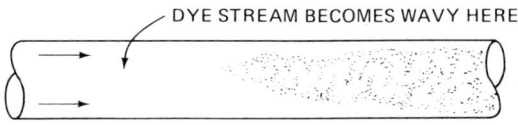

FIG. 9-3. Onset of turbulent flow.

9-4 REYNOLDS NUMBER

In his work Reynolds was able to develop a mathematical method to further define laminar and turbulent flow conditions. This mathematical approach involves the determination of a dimensionless number called the *Reynolds number* (N_R). In considering laminar and turbulent flow, it is reasonable to conclude that the factors listed below influence the type of flow:

1 / **Velocity of the fluid.**
2 / **Viscosity and density of the fluid.**
3 / **Diameter of the pipe (a circular cross section is assumed).**

In his work Reynolds was able to prove that this is the case. The Reynolds number is then defined as

$$\text{Reynolds number } (N_R) = \frac{(\text{density})(\text{velocity})(\text{diameter})}{\text{absolute viscosity}}$$

Using the mathematical symbols previously adopted and letting d = diameter, the equation becomes

$$N_R = \frac{\rho v d}{\mu} \tag{9-3}$$

Equation (9-3) can be rearranged to another form using kinematic viscosity instead of the absolute viscosity. From Chapter Two, kinematic viscosity (ν) equals absolute viscosity divided by density, or μ/ρ. Equation (9-3) can be rewritten as $N_R = vd/(\mu/\rho)$ and ν can be substituted for the term μ/ρ. This gives

$$N_R = \frac{vd}{\nu} \qquad (9\text{-}4)$$

The Reynolds number has no units or dimensions. This can be shown by checking Eq. (9-4). The velocity v has units of ft/s, the diameter d is in ft, and the kinematic viscosity has units of ft²/s. With these units substituted into the right-hand side of the equation, it becomes

$$\frac{(\text{ft/s})(\text{ft})}{(\text{ft}^2/\text{s})} = \frac{\text{ft}^2/\text{s}}{\text{ft}^2/\text{s}}$$

All units cancel. The number then has no dimensions and is described as dimensionless.

9-5 FLOW TYPE AND REYNOLDS NUMBER

The Reynolds number has been related to the type of flow (laminar or turbulent) in pipes by determining a critical value which separates laminar from turbulent flow. The research indicates that this critical value of Reynolds number is not precise and a range of critical values exists. For the lower critical value of Reynolds number, the following statement applies: *For pipes, laminar flow exists when the average Reynolds number is 2000 or less.* For turbulent flow in pipes, an upper critical value of Reynolds number is used: *For pipes, turbulent flow exists when the average Reynolds number is 3000 or greater.* The range between 2000 and 3000, then, is an area of uncertainty where the flow type cannot be determined. Various factors that affect the flow are pipe roughness, disturbances, and entrance conditions into the pipe.

EXAMPLE 1

Five cubic feet per second of water at 90° F is flowing in a 2-in.-diameter pipe. Determine if the flow is laminar or turbulent.

Solution

From Eq. (8-2), $Q = Av$. Solving for the velocity v,

$$v = \frac{Q}{A} = \frac{5}{(\pi/4)(\tfrac{2}{12})^2} = \frac{(5)(4)(36)}{\pi}$$

$$= 229 \; \frac{\text{ft}}{\text{s}}$$

Sect. 9-6 / DETERMINATION OF FRICTION LOSS

To determine the Reynolds number, Eq. (9-4) is used:

$$N_R = \frac{vd}{\nu} \qquad (9\text{-}4)$$

The kinematic viscosity of water at 90° F is determined from Table III in the Appendix. Its value is 0.8×10^{-5} ft^2/s. With the diameter in feet, the units of all the known terms are consistent, and they can be substituted directly into the equation:

$$N_R = \frac{(229)(2/12)}{(0.8)(10^{-5})} = \frac{(229)(10^5)}{(0.8)(6)}$$
$$= 47.7 \times 10^5$$

This is a high Reynolds number and the flow is turbulent.

EXAMPLE 2

Medium fuel oil with a kinematic viscosity of 4.12×10^{-5} ft^2/s is to be piped through a 6-in.-diameter pipe. What is the maximum velocity that can be used if the flow is to remain laminar?

Solution

The lower critical Reynolds number of 2000 applies in this case. Using Eq. (9-4),

$$N_R = \frac{vd}{\nu}$$

$$v = \frac{N_R \nu}{d} = \frac{(2000)(4.12 \times 10^{-5})}{\frac{6}{12}}$$

$$= 0.16 \, \frac{\text{ft}}{\text{s}}$$

9-6 DETERMINATION OF FRICTION LOSS

The basic equation for determining friction loss in a pipe is called the *Darcy equation*. It is an empirical equation, meaning that it is based on the results of tests rather than on theory. Partly because of this and also because of the difficulty in knowing exactly all of the conditions affecting friction, friction-loss determination is somewhat imprecise and approximate. Its accuracy is sufficient for most engineering calculations, however.

The Darcy equation for friction loss is as follows:

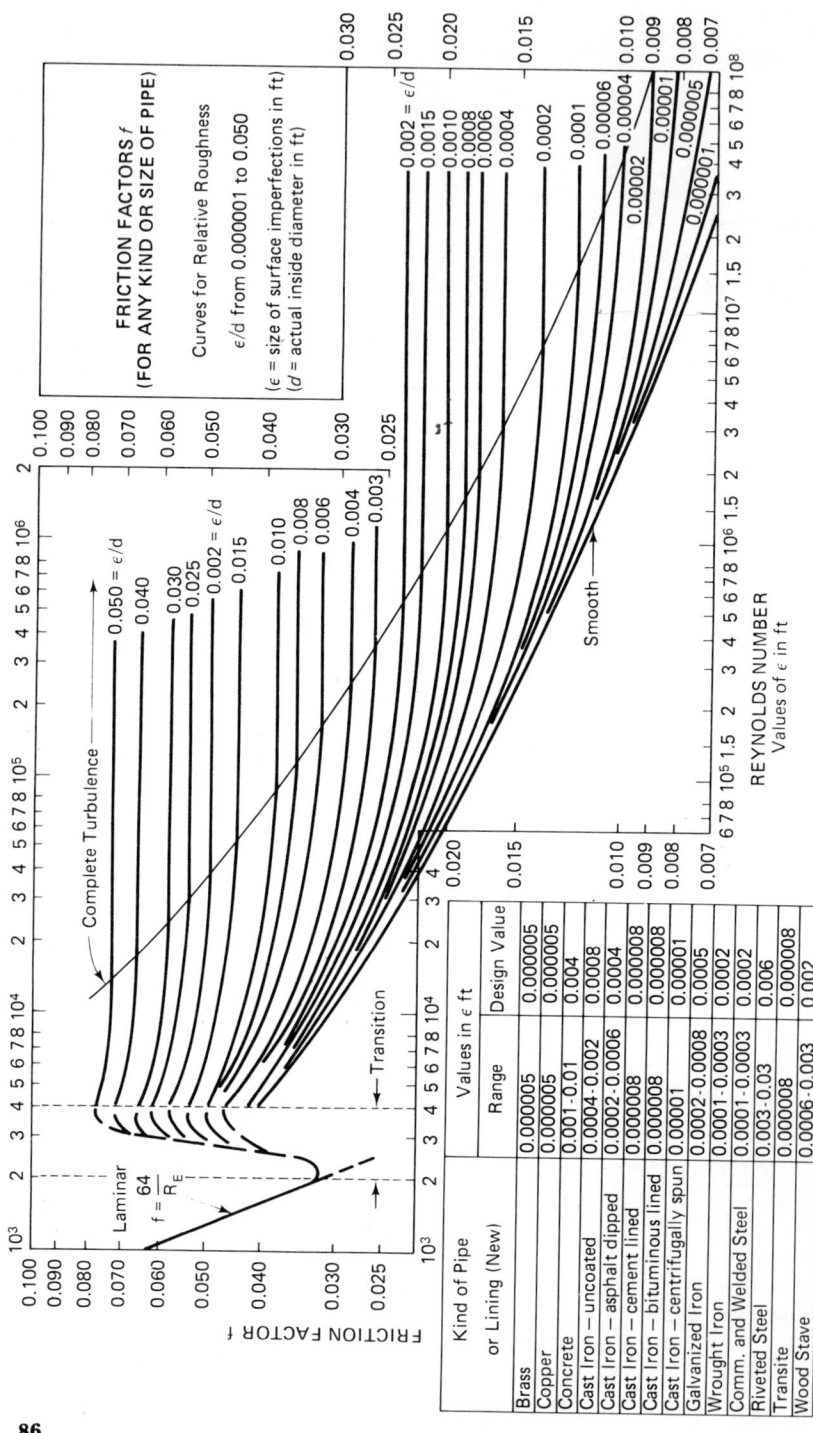

FIG. 9-4. Moody's chart and roughness value. (ASME Transactions 66, 8 Nov. 1944, *reproduced by permission of* American Society of Mechanical Engineers.)

$$h_L = f \frac{L}{d} \frac{v^2}{2g} \qquad (9\text{-}5)$$

where h_L = friction loss in feet of the liquid used
L = length of pipe, in ft
d = pipe diameter, ft
v = velocity, ft/s
f = a friction factor depending on the Reynolds number, the roughness of the pipe, and the pipe diameter

Determining the value to be used for f in Eq. (9-5) requires consideration of the flow type, whether laminar or turbulent. If the flow is laminar, f can be represented by a simple expression, as follows:

$$f = \frac{64}{N_R} \qquad (9\text{-}6)$$

Equation (9-6) is the result of studies of Poiseuille, a French scientist.

If the flow is turbulent, the value of f is determined by the use of a combination or combinations of equations developed by various researchers. These are combined in the chart developed by L. F. Moody ("Friction Factors for Pipe Flow," *American Society of Mechanical Engineers Transactions*, vol. 66, Nov. 1944). This is reproduced in Fig. 9-4.

9-7   USE OF MOODY'S CHART

Moody's chart plots values of f for turbulent flow as well as the laminar flow values of f. It is most useful for determining f under turbulent flow, since Eq. (9-6) is in such a simple form that f can be readily computed.

Pipe roughness affects friction loss under turbulent flow conditions. If the absolute roughness of the pipe surface is defined by the average height epsilon (ϵ) of the irregularities as shown in Fig. 9-5, the ratio ϵ/d

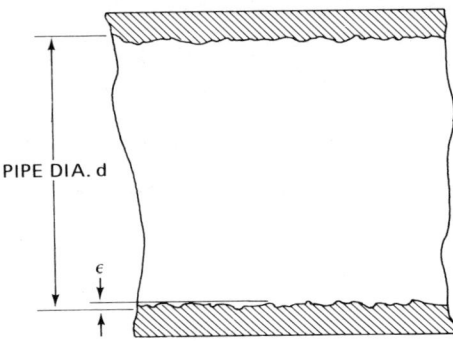

FIG. 9-5.   Pipe roughness.

(d is the pipe diameter) is called the *relative roughness of the surface*. This ratio is a factor in the friction loss and is plotted on the right-hand side of Moody's chart. The Reynolds number is also a factor and is plotted at the bottom of the chart.

The selection of the values to be used for ϵ is an estimate frequently, since the exact pipe condition may not be known. Figure 9-4 contains a table with values of ϵ for various materials. This table is used as the basis for estimating the value of ϵ.

The steps in using the chart are as follows (and see Fig. 9-6):

1 / **Compute the Reynolds number.**
2 / **Select a value of ϵ to be used for the roughness.**
3 / **Compute ϵ/d.**
4 / **Using the ϵ/d value calculated, enter the chart at the right, and approximate a curve for this value as shown in Fig. 9-6.**
5 / **With the calculated Reynolds number, enter the chart from the bottom and extend the number vertically until it intersects the ϵ/d curve.**
6 / **From this point of intersection move horizontally to the left and read the friction factor f from the left-hand column. The friction factor f is dimensionless.**

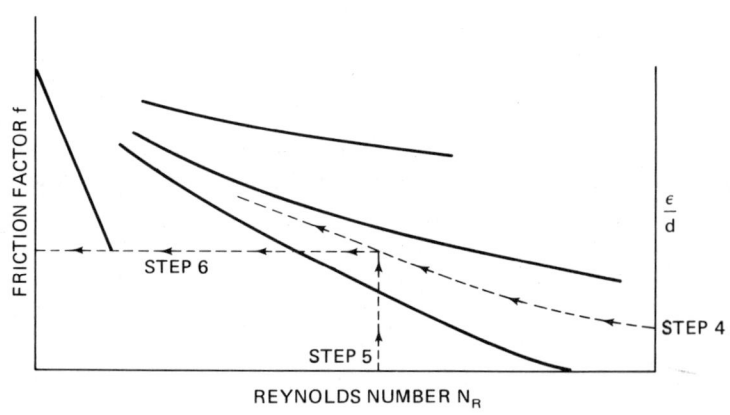

FIG. 9-6.

Once the value of f is determined, it is substituted into Eq. (9-5). Equation (9-5) can then be solved for the head loss (or other unknown if the head loss is known).

EXAMPLE 1

Water flows in a concrete pipe at a Reynolds number of 5.5×10^5. Its velocity is 10 ft/s and the pipe's diameter is 1 ft. Determine the head loss per 100 ft of pipe.

Solution

Referring to Fig. 9-4, note that the absolute roughness of concrete varies from 0.01 to 0.001. Since no mention is made of the degree of roughness of the concrete in the problem, it is reasonable to assume that an average roughness figure may be used. Epsilon (ϵ) may then be assumed to be 0.005. Then

$$\frac{\epsilon}{d} = \frac{0.005}{1} = 0.005 \qquad \text{(to make the units consistent, } d \text{ must be in ft)}$$

Using $\epsilon/d = 0.005$ and $N_R = 5.5 \times 10^5$, use Fig. 9-4 to determine the value of f as 0.031. Substituting these values, Eq. (9-5) becomes

$$h_L = 0.031 \; \frac{100}{1} \; \frac{(10)^2}{(2)(32.2)} \quad (g = 32.2 \text{ ft/sec}^2)$$
$$= 4.8 \text{ ft of water per 100 ft of pipe}$$

Converted to psi, this becomes $(4.8)(0.433) = 2.1$ psi.

EXAMPLE 2

Oil with a specific gravity of 0.855 flows as shown in Fig. 9-7 in 12-in.-diameter steel pipe from A to B. At B a 2-in. steel pipe is connected and the oil flows through this pipe to C. Velocities, viscosity, and

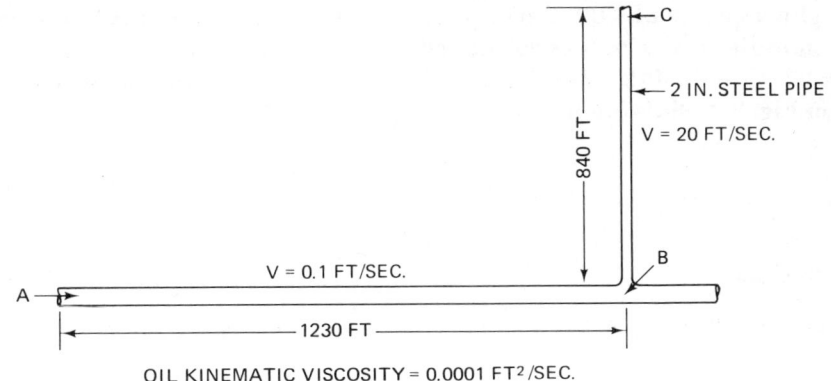

FIG. 9-7.

distances are shown in the sketch. Determine (a) the friction loss from A to B and (b) the friction loss from B to C.

Solution

The Reynolds number for the two segments of pipe are required and calculated as follows, using Eq. (9-4). From A to B,

$$N_R = \frac{vd}{\nu} = \frac{0.1 \text{ ft/s } (1 \text{ ft})}{0.0001 \text{ ft}^2/\text{s}}$$

$$= 1000$$

Flow here is laminar since N_R is less than 2000. From B to C,

$$N_R = \frac{20 \text{ ft/s } (\frac{2}{12}) \text{ ft}}{0.0001 \text{ ft}^2/\text{s}} = 3.3 \times 10^4$$

Flow from B to C is turbulent. Since the flow from A to B is laminar, Eq. (9-6) applies for determining the friction factor. This becomes

$$f = \frac{64}{N_R} = \frac{64}{1000} = 0.064$$

Using Eq. (9-5) and $f = 0.064$,

$$h_L = f \frac{L}{d} \frac{v^2}{2g} = 0.064 \left[\frac{1230 \text{ ft}}{1 \text{ ft}} \frac{(0.1 \text{ ft/s})^2}{(2)(32.2 \text{ ft/s}^2)} \right]$$

$$= 0.012 \text{ ft of oil}$$

The head loss in this segment is insignificant and can be neglected.

For the segment from B to C, Fig. 9-4 is used to determine f. A roughness value of 0.0002 seems reasonable to use, since it is about in the middle of the values tabulated for steel. The ratio ϵ/d then is $0.0002/\frac{2}{12} = 0.0012$. With $N_R = 3.3 \times 10^4$, f is found to be 0.026 from Fig. 9-4. Solving for h_L,

$$h_L = 0.026 \frac{840}{\frac{2}{12}} \frac{(20)^2}{(2)(32.2)}$$

$$= 814 \text{ ft}$$

To convert this to psi, the specific gravity must be used. Then

$$814 \text{ ft} (0.855) \; 0.433 \frac{\text{psi}}{\text{ft}} = 301 \text{ psi}$$

QUESTIONS AND PROBLEMS

9-1.—Using Fig. 9-4, determine the Reynolds number required to give a friction factor of 0.030 with $\epsilon/d = 0.004$.

9-2.—With $f = 0.030$, what Reynolds number is required to give laminar flow?

9-3.—Water flows in a 1-in. copper tube at a Reynolds number of 1×10^7. What is the friction factor?

9-4.—Twenty gallons per minute of water flows through 200 ft of 1-in. commercial steel pipe. The water viscosity is 0.000015 ft^2/s. Calculate the head loss due to friction.

9-5.—One thousand gallons per minute of water flows in the system in Fig. 9-8. A constant head of 30 ft is kept at A and the pipe is 6-in.-inside-diameter commercial steel. What is the pressure at B? (*Hint:* Calculate the friction loss, then use Bernoulli's equation.)

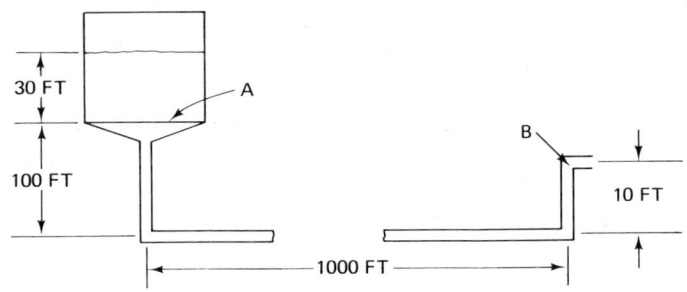

FIG. 9-8.

9-6.—A test to determine the friction loss in the 1000-ft length of pipe shown in Fig. 9-9 is made by locating pressure gages at A and B. The pipe is horizontal and the liquid is medium fuel oil. With other conditions as shown, calculate the friction factor f. (*Note:* Under these conditions, the friction loss is $P_A - P_B$. Use Eq. (9-5) and Fig. 9-4 for f; Table III has viscosity values.)

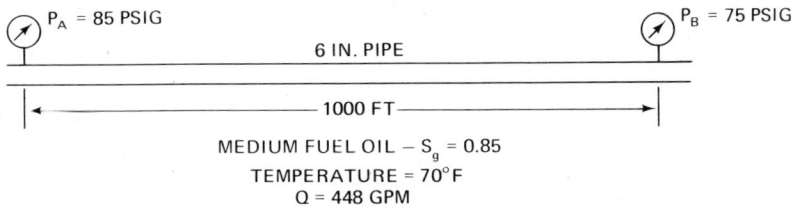

FIG. 9-9.

9-7.—Determine the absolute roughness of the pipe in Problem 9-6.

9-8.—Determine the Reynolds number for crude oil with a specific gravity of 0.855 flowing in a 4-in.-diameter pipe at a velocity of 5 ft/s at

    a—50°F.

    b—90° F.

9-9.—Kerosene flows in 1-in. wrought iron pipe at a velocity of 2 ft/s and at a temperature of 70° F. Determine the pressure drop due to friction in 50 ft of the pipe.

chapter ten

FLUID DYNAMICS— COMPRESSIBLE FLUIDS

10-1 INTRODUCTION

The fundamental theory of fluid dynamics applies to compressible as well as incompressible fluids. Its application to compressible fluids is more complicated than it is for incompressible fluids. The reason for this is the fact that the fluid is compressible and frequently changes its volume for a given amount of mass. This change in volume can be imagined by examining Fig. 10-1. An imaginary volume of gas is changed from V_1 to V_2 as the gas flows from a small-diameter to a large-diameter pipe. Because of the lesser downstream pressure at point 2, the volume V_2 is greater than V_1.

Although the volume changes, the mass of this volume has to remain constant. Thus, a basis for analyzing compressible fluid dynamics is the *mass* rate of flow. Since weight can be used as well as mass, the term lb/s (or an equivalent) is used to represent a weight flow rate.

A simplified method of analysis for dealing with compressible fluid flow allows us to use the perfect-gas law:

$$PV_s = RT \qquad (4\text{-}3)$$

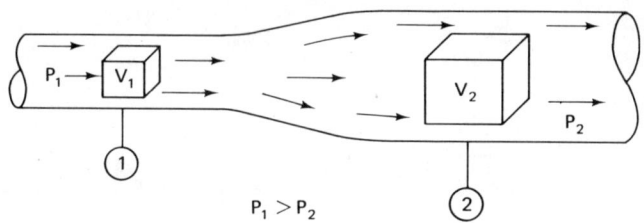

FIG. 10-1. Change in volume of a mass of gas.

Use of the perfect-gas law can be justified if the gas is of light weight and follows the perfect-gas law. Also, the velocity of the gas must be high enough so that there are reasonably large changes in pressure, temperature, and density. If these conditions are met, the thermodynamic properties and processes described in Chapters Four and Five can be applied.

10-2 GAS VOLUME AT STANDARD CONDITIONS

A commonly used volumetric flow rate of a gas (frequently air) is that described by the term "standard cubic feet per minute" abbreviated "scfm." This describes the volume of the gas flowing at standard conditions of pressure and temperature:

> **Standard Conditions: atmospheric pressure of 14.7 psia, temperature of 68°F**

A flow rate of 25 scfm of air means 25 ft³ of air as measured at 14.7 psia and 68°F is flowing each minute.

Note that the volume used is referenced to specific pressure and temperature conditions.

10-3 EQUATION OF CONTINUITY— COMPRESSIBLE FLUIDS

The principle of conservation of mass, as described in Chapter Eight, states that the total quantity of mass in a system remains constant as long as no other mass is added to or taken away from the system. This principle applies for compressible fluids. In Fig. 10-2, the same mass of the fluid flows past point 2 as past point 1. Therefore, the equation of continuity [Eq. (8-1)], in the form which expresses a *mass* or *weight* flow rate, applies to compressible fluids. Letting G = weight flow rate and rewriting the equation, it becomes

$$G = W_{s1} A_1 v_1 = W_{s2} A_2 v_2 \qquad (10\text{-}1)$$

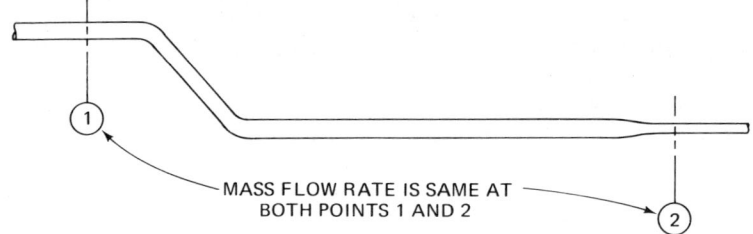

FIG. 10-2. Mass flow of a compressible fluid.

If W_s is in the units of lb/ft³, A is in units of ft², and v is in units of ft/s, then G is in units of lb/s. Subscripts 1 and 2 indicate two different locations, 1 and 2. The specific weight and velocity terms in the equation must be at the conditions and locations of the system in question.

10-4 BERNOULLI'S EQUATION FOR COMPRESSIBLE FLUIDS

The principle of conservation of energy applies for compressible fluids just as does the principle of conservation of mass. It is logical to assume, then, that Bernoulli's equation may be used for compressible fluids. With no energy added to the system and assuming no energy lost due to friction, Bernoulli's equation was described in Chapter Eight in the following manner:

$$\begin{bmatrix}\text{press.}\\\text{head}\end{bmatrix}_1 + \begin{bmatrix}\text{vel.}\\\text{head}\end{bmatrix}_1 + \begin{bmatrix}\text{elev.}\\\text{head}\end{bmatrix}_1 = \begin{bmatrix}\text{press.}\\\text{head}\end{bmatrix}_2 + \begin{bmatrix}\text{vel.}\\\text{head}\end{bmatrix}_2 + \begin{bmatrix}\text{elev.}\\\text{head}\end{bmatrix}_2$$

or, in mathematical terms,

$$\frac{P_1}{W_s} + \frac{v_1^2}{2g} + Z_1 = \frac{P_2}{W_s} + \frac{v_2^2}{2g} + Z_2 \qquad (8\text{-}5)$$

Under most conditions the elevation head for a gas is insignificant and is neglected. This is shown in the calculations in Section 6-5, where the elevation head represents less than 0.25 percent of the total pressure head involved in the example. The elevation head Z then becomes 0 in Eq. (8-5).

The specific weight, W_s, is a variable when the fluid is compressible. The equation now becomes, with $Z = 0$,

$$\frac{P_1}{W_{s1}} + \frac{v_1^2}{2g} = \frac{P_2}{W_{s2}} + \frac{v_2^2}{2g} \qquad (10\text{-}2)$$

The process of compressible fluid flow is considered to be an adiabatic process. This assumption is made because the rapidity with which changes are made in the fluid prevent any appreciable heat transfer from taking place. In Chapter Five, the adiabatic process is defined as one in which no heat transfer takes place. The equation for this process is

$$P_1 V_1^{1.4} = P_2 V_2^{1.4} \tag{5-3}$$

where 1.4 is the value of the exponent, n, for air. Since $W_s = 1/V_s$, there are now four equations which can be related. They are:

Eq. (10-1): $\quad G = W_s A v$

Eq. (10-2): $\quad \dfrac{P_1}{W_{s1}} + \dfrac{v_1^2}{2g} = \dfrac{P_2}{W_{s2}} + \dfrac{v_2^2}{2g}$

Eq. (5-3): $\quad P_1 V_1^{1.4} = P_2 V_2^{1.4}$

Eq. (4-3): $\quad PV_s = RT$

The last three have been combined to give several modified forms of Bernoulli's equation for compressible flow. Since the derivation of these forms is somewhat complicated, it is not given here. Three forms of this equation are now given:

$$\dfrac{v_2^2 - v_1^2}{2g} = \dfrac{P_1}{W_{s1}} \dfrac{n}{n-1} \left[1 - \left(\dfrac{P_2}{P_1}\right)^{(n-1)/n}\right] \tag{10-3}$$

$$\dfrac{v_2^2 - v_1^2}{2g} = \dfrac{P_2}{W_{s2}} \dfrac{n}{n-1} \left[\left(\dfrac{P_1}{P_2}\right)^{(n-1)/n} - 1\right] \tag{10-4}$$

$$\dfrac{v_2^2 - v_1^2}{2g} = \dfrac{Rn}{n-1}(T_1 - T_2) \tag{10-5}$$

Pressure and temperature terms in these equations are absolute values.

EXAMPLE 1

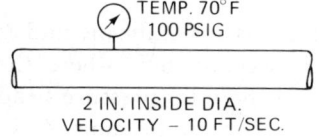

FIG. 10-3.

Air flows in the pipe in Fig. 10-3. The velocity is measured by a pitot tube and found to be 10 ft/s. Determine the flow volume in standard cubic feet per minute (scfm).

Solution

The equation of continuity in terms of volume ($Q = Av$) can be used to determine the volume at the conditions in the pipe. This volume

Sect. 10-4 / BERNOULLI'S EQUATION FOR COMPRESSIBLE FLUIDS

can then be converted to the volume at standard conditions.

$$Q = \frac{\pi}{4}\left(\frac{2}{12}\right)^2 (10) = 0.218 \; \frac{\text{ft}^3}{\text{sec}}$$

$$= (0.218)(60) = 13.1 \; \frac{\text{ft}^3}{\text{min}}$$

The volume of 13.1 ft³ is that which exists at 100 psig pressure and 70°F temperature. This can be changed to standard conditions by using the perfect-gas law in the form $PV = WRT$ [Eq. (4-4)]. Using subscripts 1 and 2, $P_1 V_1 = WRT_1$ and $P_2 V_2 = WRT_2$, divide one equation by the other.

$$\frac{P_1 V_1}{P_2 V_2} = \frac{WRT_1}{WRT_2}$$

The W and R terms cancel, and

$$V_2 = \frac{T_2}{T_1}\left(\frac{P_1}{P_2}\right) V_1$$

Pressure and temperatures are absolute values.

$$T_1 = 70 + 460 = 530°R$$
$$T_2 = 68 + 460 = 528°R$$
$$P_1 = 100 + 14.7 = 114.7 \text{ psia}$$
$$P_2 = 14.7 \text{ psia}$$

These values are substituted into the equation and

$$V_2 = \frac{528}{530}\left(\frac{114.7}{14.7}\right)(13.1)$$

$$= 102 \text{ scfm}$$

EXAMPLE 2

Air flows in the duct in Fig. 10-4. The flow conditions at the two points are as shown. Determine the velocities of flow at the two points and the flow rate in lb/s.

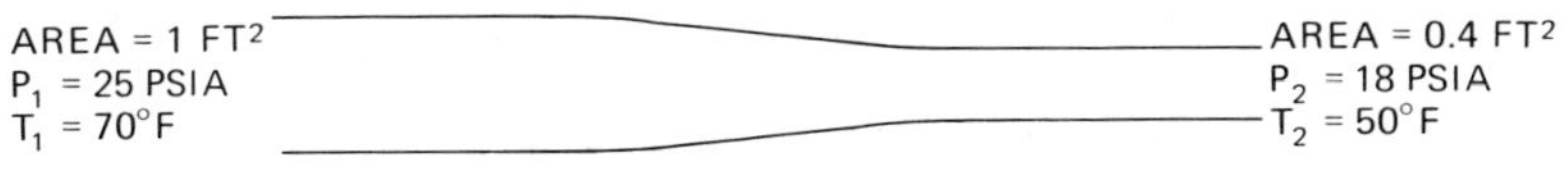

```
AREA = 1 FT²                              AREA = 0.4 FT²
P₁ = 25 PSIA                              P₂ = 18 PSIA
T₁ = 70°F                                 T₂ = 50°F
```

FIG. 10-4.

Solution

Forms of the perfect-gas law, the equation of continuity, and Bernoulli's equation are used in the solution. The specific weight at each point is determined from Eq. (4-3).

$$PV_s = RT \qquad V_s = \frac{1}{W_s}$$

$$P\left(\frac{1}{W_s}\right) = RT \qquad W_s = \frac{P}{RT}$$

At point 1,

$$W_{s1} = \frac{(25)(144)}{(53.3)(70 + 460)}$$

$$= 0.1274 \; \frac{\text{lb}}{\text{ft}^3}$$

At point 2,

$$W_{s2} = \frac{(18)(144)}{(53.3)(50 + 460)}$$

$$= 0.0953 \; \frac{\text{lb}}{\text{ft}^3}$$

The equation of continuity, Eq. (10-1), can be used now to obtain a relationship between the velocities at the two points.

$$G = W_{s1} A_1 v_1 = W_{s2} A_2 v_2 \tag{10-1}$$

and

$$W_{s1} A_1 v_1 = W_{s2} A_2 v_2$$

Substituting,

$$(0.1274)(1) v_1 = (0.0953)(0.4) v_2$$

$$v_1 = 0.299 v_2$$

Equation (10-3), (10-4), or (10-5) can be used now to determine the velocities. Since Eq. (10-5) appears less complicated, this one is used.

$$\frac{v_2^2 - v_1^2}{2g} = \frac{Rn}{n - 1} (T_1 - T_2)$$

$$\frac{v_2^2 - v_1^2}{(2)(32.2)} = \frac{(53.3)(1.4)}{1.4 - 1} (530 - 510)$$

$$v_2^2 - v_1^2 = 240{,}000$$

Sect. 10-4 / BERNOULLI'S EQUATION FOR COMPRESSIBLE FLUIDS

Substituting $0.299 v_2$ for v_1,

$$v_2^2 - (0.299 v_2)^2 = 240,000$$

$$v_2 = 513 \ \frac{\text{ft}}{\text{s}} \quad \text{and} \quad v_1 = (0.299)(513) = 153 \ \frac{\text{ft}}{\text{s}}$$

The flow rate can now be determined from Eq. (10-1):

$$G = W_{s1} A_1 v_1 = (0.1274)(1)(153)$$
$$= 19.5 \ \frac{\text{lb}}{\text{s}}$$

EXAMPLE 3

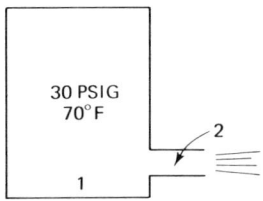

FIG. 10-5.

Air is contained in the tank as shown in Fig. 10-5 at a pressure of 30 psig and 70° F. It is exhausted to atmosphere through a round nozzle with a cross-sectional area of 0.4 ft.² Calculate the flow rate in lb/s. Assume that the process is adiabatic.

Solution

Conditions inside the tank are referred to as point 1 and conditions inside the nozzle as point 2. If the velocity of flow and the specific weight at point 2 can be determined, the flow rate can be found from the equation of continuity. The pressure at point 2 is taken to be atmospheric.

$$P\left(\frac{1}{W_{s1}}\right) = RT_1 \quad \text{and} \quad W_{s1} = \frac{P_1}{RT_1}$$

$$W_{s1} = \frac{(30 + 14.7)(144)}{(53.3)(70 + 460)}$$

$$= 0.228 \ \frac{\text{lb}}{\text{ft}^3}$$

Equation (10-3) can be used to solve for the velocity at point 2. The velocity at point 1 is 0 and the pressures and specific weight at point 1 are known.

$$\frac{v_2^2 - v_1^2}{2g} = \frac{P_1}{W_{s1}} \frac{n}{n-1} \left[1 - \left(\frac{P_2}{P_1}\right)^{(n-1)/n} \right]$$

Substituting,

$$\frac{v_2^2 - 0}{(2)(32.2)} = \frac{(30 + 14.7)(144)}{0.228} \frac{1.4}{1.4 - 1} \left[1 - \left(\frac{14.7}{30 + 14.7}\right)^{(1.4-1)/1.4}\right]$$

$$v_2^2 = (64.4)(144) \frac{44.7}{0.228} \frac{1.4}{0.4} \left[1 - \left(\frac{14.7}{44.7}\right)^{0.4/1.4}\right]$$

$$= (44,200)(144)[1 - (0.329)^{0.286}] = (6.36 \times 10^6)(1 - 0.727)$$

$$v_2 = 1.32 \times 10^3 \frac{\text{ft}^3}{\text{s}}$$

Since the process is adiabatic, Eq. (5-3) applies.

$$P_1 V_1^{1.4} = P_2 V_2^{1.4} \tag{5-3}$$

Changing this to use the specific weight,

$$P_1 \left(\frac{1}{W_{s1}}\right)^{1.4} = P_2 \left(\frac{1}{W_{s2}}\right)^{1.4} \quad \text{and} \quad \frac{P_1}{W_{s1}^{1.4}} = \frac{P_2}{W_{s2}^{1.4}}$$

Solving for W_{s2},

$$W_{s2}^{1.4} = \frac{P_2}{P_1} (W_{s1})^{1.4}$$

$$W_{s2} = \left(\frac{P_2}{P_1}\right)^{1/1.4} W_{s1}$$

Substituting,

$$W_{s2} = \left(\frac{14.7}{44.7}\right)^{1/1.4} (0.228) = (0.328)^{0.714} (0.228)$$

$$= (0.45)(0.228) = 0.1025 \frac{\text{lb}}{\text{ft}^3}$$

The flow rate now is

$$G = W_{s2} A_2 v_2 = (0.1025)(0.4)(1320)$$

$$= 54.1 \frac{\text{lb}}{\text{s}}$$

10-5 MACH NUMBER

An important term in aerodynamics and certain other fields where air velocity in a system is critical is the *Mach number*. The Mach number is the ratio of the velocity of the observed air to that of the acoustic (sonic) velocity. The acoustic velocity is the velocity of sound in air at 59° F and is 1120 ft/s. For ratios less then 1, the flow is termed *subsonic*. For ratios greater than 1, the flow is *supersonic*.

QUESTIONS AND PROBLEMS

10-1.—Determine the scfm of air equivalent to 25 ft^3/min at 50° F and 35 psig.

10-2.— The flowmeter in Fig. 10-6 is calibrated to read directly in standard cubic feet per minute with air flowing in it. Determine the lb of air flowing per second when it indicates 35 scfm.

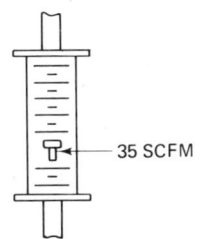

FIG. 10-6.

10-3.— Determine the flow rate in scfm and lb/s of the air flowing in the pipe in Fig. 10-7.

FIG. 10-7.

10-4.—Air flows in the duct in Fig. 10-8. Determine the velocities of flow at the two points and the flow rate in lb/s.

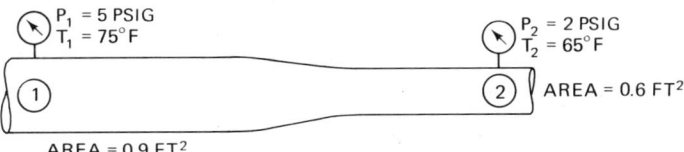

FIG. 10-8.

10-5.—

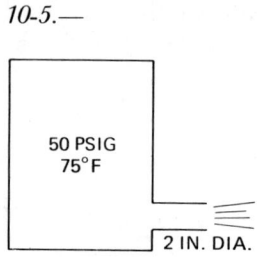

FIG. 10-9.

Air pressure is maintained at 50 psig in the tank as shown in Fig. 10-9. The air exhausts to atmosphere through the nozzle at the bottom. Determine the flow rate in lb/s.

chapter eleven

FRICTION LOSS—AIR

11-1 INTRODUCTION

Fluid friction occurs in a system using compressible fluids just as it does with incompressible fluids. Under constant-flow conditions, the fluid friction results in a pressure drop, as shown in Fig. 11-1. Note that the pipe cross section is constant here, so velocity and corresponding pressure changes do not occur as a result of a change in section.

Calculation of pressure drop due to friction is imprecise for compressible fluids just as it is for incompressible fluids. It depends on empirical equations and methods to a greater extent than does the calculation for incompressible fluids. Since the gas most used in industry is air, most of the work done and most of the data available are on air. Because of this, only methods pertaining to air are presented here.

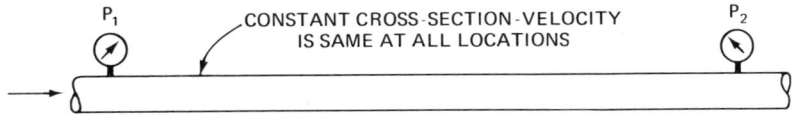

FIG. 11-1.   Pressure drop ($P_1 - P_2$) caused by friction.

103

Table 11-1. Harris Formula Flow Factor F For Air in Standard Schedule 40 Steel Pipe.*

Q scfm	Nominal diameter, in.										
	$\frac{1}{2}$	$\frac{3}{4}$	1	$1\frac{1}{4}$	$1\frac{1}{2}$	2	$2\frac{1}{2}$	3			
5	8.9	2.0	0.5	—	—	—	—	—			
10	35.4	8.0	2.2	0.5	—	—	—	—			
15	79.7	17.9	4.9	1.1	—	—	—	—			
20	142	31.8	8.7	2.0	0.9	—	—	—			
25	221	49.7	13.6	3.2	1.4	—	—	—			
30	318	71	19.6	4.5	2.0	0.7	—	—			
35	434	97.5	26.6	6.2	2.7	1.1	—	—			
40	567	127	34.8	8.1	3.6	1.4	—	—			
45	716	161	44.0	10.2	4.5	1.9	—	—			
50	885	199	54.4	12.6	5.6	2.4	1.2	—			
60	—	286	78.3	18.2	8.0	2.9	1.5	—			
70	—	390	106.6	24.7	10.9	4.2	2.2	—			
80	—	510	139.2	32.3	14.3	5.7	2.9	—			
90	—	645	176.2	40.9	18.1	7.5	3.8	—			
100	—	796	217.4	50.5	22.3	9.5	4.8	—			
110	—	963	263	61.1	27.0	11.7	6.0	—			
120	—	—	318	72.7	32.2	14.1	7.2	—			
130	—	—	369	85.3	37.8	16.8	8.6	1.2			
140	—	—	426	98.9	43.8	19.7	10.1	1.4			
150	—	—	490	113.6	50.3	22.9	11.7	1.6			
160	—	—	570	120.3	57.2	26.3	13.4	1.9			
170	—	—	628	145.8	64.6	29.9	15.3	1.9			
180	—	—	705	163.3	72.6	33.7	17.6	2.1			
190	—	—	785	177	80.7	37.9	19.4	2.4			
200	—	—	870	202	89.4	42.2	21.5	2.6			
220	—	—	—	244	108.2	46.7	23.9	2.9			
240	—	—	—	291	128.7	56.5	28.9	3.5			
260	—	—	—	341	151	67.3	34.4	4.2			
280	—	—	—	395	175	79.0	40.3	4.9			
300	—	—	—	454	201	91.6	46.8	5.7			

Q scfm	Nominal diameter, in.							
	2	$2\frac{1}{2}$	3	$3\frac{1}{2}$	4	$4\frac{1}{2}$	5	6	8	10
320	61.1	23.8	7.5	3.5	—	2.0	—	—	—	—
340	69.0	26.8	8.4	3.9	—	2.2	—	—	—	—
360	77.3	30.1	9.5	4.4	2.0	—	—	—	—	—
380	86.1	33.5	10.5	4.9	2.2	—	—	—	—	—
400	94.7	37.1	11.7	5.4	2.5	—	—	—	—	—
420	105.2	40.9	12.9	6.0	2.7	—	—	—	—	—
440	115.5	44.9	14.1	6.6	3.1	—	—	—	—	—

Q scfm	Nominal diameter, in.									
	2	$2\frac{1}{2}$	3	$3\frac{1}{2}$	4	$4\frac{1}{2}$	5	6	8	10
460	125.6	48.8	15.4	7.1	3.7	2.0	—	—	—	—
480	137.6	53.4	16.8	7.8	4.0	2.2	—	—	—	—
500	150.0	58.0	18.3	8.5	4.3	2.4	—	—	—	—
525	165.0	64.2	20.2	9.4	4.8	2.6	—	—	—	—
550	181.5	70.2	22.1	10.2	5.2	2.9	—	—	—	—
575	197	76.7	24.2	11.2	5.7	3.1	—	—	—	—
600	215	83.5	26.3	12.2	6.2	3.4	—	—	—	—
625	233	92.7	28.5	13.2	6.8	3.7	—	—	—	—
650	253	98.0	30.9	14.3	7.3	4.0	2.2	—	—	—
675	272	105.7	33.3	15.4	7.9	4.3	2.4	—	—	—
700	294	113.7	35.8	16.6	8.5	4.6	2.6	—	—	—
750	337	130.5	41.1	19.0	9.7	5.3	2.9	—	—	—
800	382	148.4	46.7	21.7	11.1	6.1	3.3	—	—	—
850	433	168	52.8	24.4	12.5	6.8	3.8	—	—	—
900	468	188	59.1	27.4	14.0	7.7	4.2	—	—	—
950	541	209.4	65.9	30.5	15.7	8.6	4.7	—	—	—
1000	600	232.0	73.0	33.8	17.3	9.5	5.2	1.9	—	—
1050	658	256	80.5	37.8	19.1	10.4	5.8	2.1	—	—
1100	723	280.6	88.4	40.9	21.0	11.5	6.3	2.4	—	—
1150	790	306.8	96.6	44.7	22.9	12.5	6.9	2.6	—	—
1200	850	344.0	105.2	48.8	25.0	13.7	7.5	3.3	—	—
1300	—	392.0	123.4	57.2	29.3	16.0	8.8	3.8	—	—
1400	—	—	—	66.3	33.9	18.6	10.2	3.8	—	—
1500	—	—	—	76.1	39.0	21.3	11.8	4.4	—	—
1600	—	—	—	86.6	44.3	24.2	13.4	5.1	—	—
1700	—	—	—	97.8	50.1	27.4	15.1	5.7	—	—
1800	—	—	—	110.0	56.1	30.7	16.9	6.4	—	—
1900	—	—	—	122	62.7	34.2	18.9	7.1	1.6	—
2000	—	—	—	135	69.3	37.9	20.9	7.8	1.8	—
2100	—	—	—	149	76.4	40.8	23.0	8.7	2.0	—
2200	—	—	—	166	83.6	45.8	25.3	9.5	2.2	—
2300	—	—	—	179	91.6	50.1	27.6	10.4	2.4	—
2400	—	—	—	195	99.8	54.6	30.1	11.3	2.6	—
2500	—	—	—	212	108.3	59.2	32.6	12.3	2.9	—
2600	—	—	—	229	117.2	64.9	35.3	13.3	3.1	—
2700	—	—	—	247	126	69.1	38.1	14.3	3.3	—
2800	—	—	—	265	136	74.3	41.0	15.4	3.6	—
2900	—	—	—	285	146	79.8	43.9	16.5	3.9	—
3000	—	—	—	305	156	85.2	47.0	17.7	4.1	—

*Courtesy Addison-Wesley Publishing Co., reproduced with permission from *Introduction to Fluid Mechanics*, by Russell W. Henke.

11-2   AIR-FLOW LOSSES IN PIPE— HARRIS FORMULA

The *Harris formula* is applicable to flow losses in Schedule 40 (standard-weight) steel pipe. Schedule 40 steel pipe is commonly used for compressed-air piping. Its properties, uses, and other characteristics are covered in more detail in later chapters.

The Harris formula is a complicated one. It is shown here with the necessary definitions of the terms:

$$P_f = \frac{0.1025 LQ^2}{rd^{5.31}} \qquad (11\text{-}1)$$

where P_f = pressure drop, psi
L = length of pipe, ft
r = compression ratio = $\dfrac{\text{absolute pressure in pipe}}{\text{atmospheric pressure}}$
Q = flow rate, scfs
d = actual internal pipe diameter, in.

To aid in using this formula, the most complicated part of the equation has been calculated for each pipe size and the results incorporated into a table. The table is shown in Table 11-1. The table is set up in the following manner:

1 / A term called the *flow factor*, F, has been calculated for 1000 ft of each pipe size. The flow factor contains the $d^{5.31}$ and 0.1025 terms of the equation.

2 / Q, the air flow quantity in scfm, and the pipe sizes are listed in the table. For each value of Q and the pipe size the flow factor F has been calculated.

To use this method, determine the compression ratio r and the flow factor F. Divide the compression ratio into the flow factor. The answer is the pressure drop for 1000 ft of the pipe size used. Stated mathematically, it is

$$P_f \text{ for 1000 ft of pipe} = \frac{F}{r} \qquad (11\text{-}2)$$

Pressure drops for lengths other than 1000 ft are proportional to the length. A 500-ft pipe length would have a pressure drop equal to one half that of the 1000-ft length, for example.

EXAMPLE

One hundred scfm of air flows in 1245 ft of 1½-in. pipe at a pressure of 80 psig. Determine the pressure drop in the pipe.

Solution

The compression ratio

$$r = \frac{80 + 14.7}{14.7} = 6.44$$

Using the chart, the flow factor F for 100 scfm and 1½-in. pipe is 22.3. Then

$$P_f' = \frac{22.3}{6.44} = 3.46 \text{ psi per 1000 ft of pipe}$$

The pressure drop for 1245 ft of pipe is

$$\frac{1245}{1000}(3.46) = 4.3 \text{ psi}$$

11-3 AIR-FLOW LOSSES IN TUBING

A method developed from National Bureau of Standards work uses the chart shown in Fig. 11-2. It is used for tubing and also for components such as valves and fittings. Only the part applicable to tubing is covered here.

Figure 11-2 contains a chart, called a *nomograph*, which is used to solve for the pressure drop. Figure 11-2 uses the following terms: Q_A, scfm; P_U, psia and also in psig; P_D, in psig; F. These terms are defined as follows:

Q_A = air flow under standard conditions, scfm
P_U = inlet or upstream pressure
P_D = outlet or downstream pressure
F = NBS flow factor

The pressure drop in the tubing, ΔP, is $P_U - P_D$.

The flow factor F may be determined from the following equations:
For short tubes (L is equal to or less than 10 times diameter)

$$F = 16.5A \tag{11-3}$$

where A is the cross-sectional area in in^2.

Sect. 11-3 / AIR-FLOW LOSSES IN TUBING

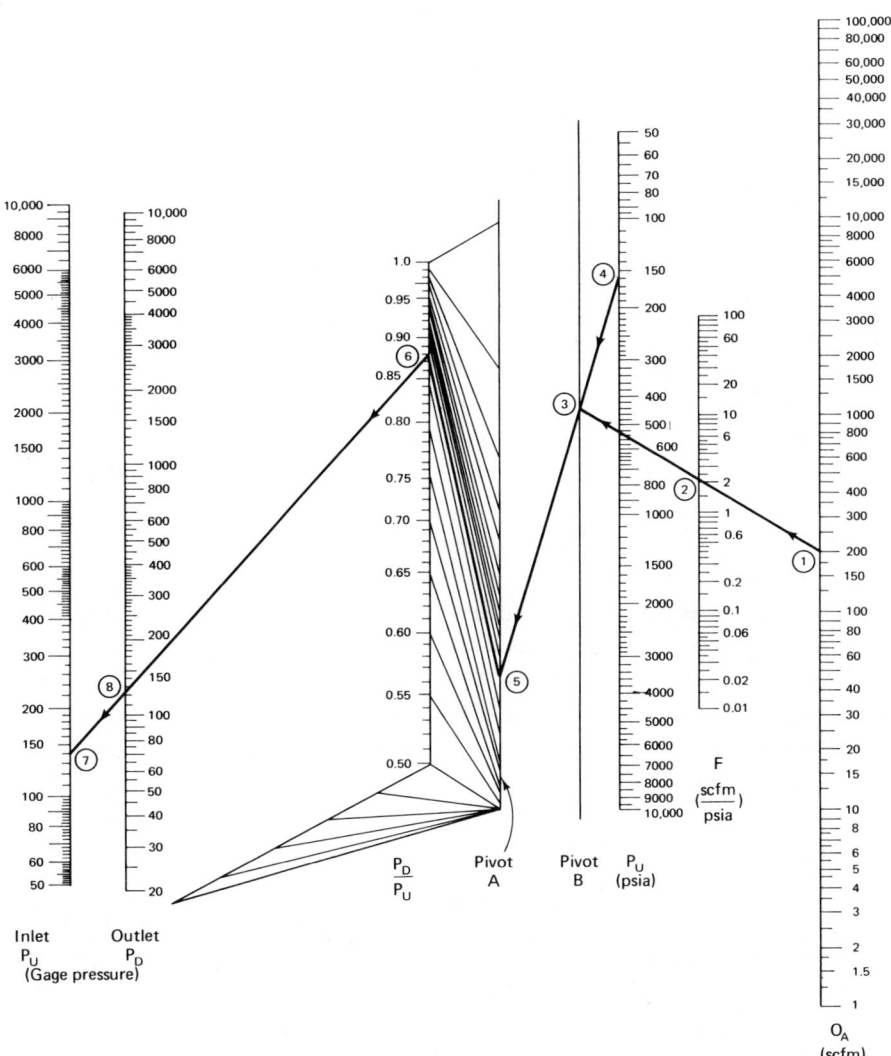

FIG. 11-2. Determination of pressure drop in tubing using National Bureau of Standards flow factor F. (*Courtesy* McGraw-Hill Book Co., *reproduced with permission*, from *Hydraulic and Pneumatic Power and Control*, ed. by Franklin D. Yeaple.)

For long tubes (L is greater than 10 times diameter)

$$F = 23.5A \sqrt{\frac{d}{fL}} \tag{11-4}$$

where d = inside diameter, in.
 A = cross-sectional area, in^2.
 L = length, in.
 f = friction factor (a typical value of 0.02 may be assumed for f)

The following procedure is used in solving problems with the nomograph.

1 / Calculate the flow factor F using the applicable equation.
2 / If the flow, Q, is not in scfm units, convert to scfm from the given temperature and pressure conditions.
3 / On the F and Q_A scales on the right of the nomograph, locate points corresponding to values for these terms.
4/ Connect the two points by a line, extending the line until it intersects pivot line B.
5 / Locate a point on the P_U absolute pressure scale corresponding to the P_U value in psia. Connect this point with the point on pivot line B and extend until it intersects pivot line A.
6 / From the point on pivot line A, draw a line to the P_D/P_U scale. This line should approximate the direction of the lines between the pivot A scale and the P_D/P_U scale.
7 / From the point on the P_D/P_U scale obtained above, draw a line to the inlet pressure P_U value in gage pressure on the left-hand scale.
8 / On the scale with the outlet pressure P_D, read the value of P_D where the line crosses.
9 / The pressure drop $\Delta P = P_U - P_D$.

EXAMPLE
Given:

$$P_U = 140 \text{ psig} = 154.7 \text{ psia}$$
$$Q_A = 200 \text{ scfm}$$
$$f = 0.02$$
$$L = 120 \text{ in.}$$
$$d = \tfrac{1}{2} \text{ in.}$$

Using Eq. (11-4),

$$F = 23.5 \frac{\pi}{4}\left(\frac{1}{2}\right)^2 \sqrt{\frac{0.5}{(0.02)(120)}}$$
$$= 2.1$$

The steps outlined previously are now followed on Fig. 11-2. Referring to Fig. 11-2, the points in the example are plotted and the sequence of the steps indicated by numbers. The final pressure, P_D, is 125 psig. The pressure drop, $\Delta P = 140 - 125 = 15$ psi.

QUESTIONS AND PROBLEMS

11-1.—Determine the pressure drop in 677 ft of 1-in. Schedule 40 pipe with 120 scfm of air flowing. The pressure in the pipe is 120 psig.

11-2.—The inlet pressure of a $1\frac{1}{2}$-in. Schedule 40 pipe carrying air is 100 psig. The length of the line is 275 ft. What is the outlet pressure if 180 scfm is flowing?

11-3.—An 8-in. length of 1-in.-inside-diameter tubing has 800 scfm of air flowing through it. With an inlet pressure of 200 psig, what is the outlet pressure?

11-4.—A 10-ft length of $\frac{3}{4}$-in.-diameter tubing has 700 scfm of air flowing through it. If the inlet pressure is 150 psig, what is the pressure drop?

11-5.—An air compressor is capable of supplying 50 scfm of air at 110 psig discharge pressure. The air is used in a shop where air tools and other equipment require 45 scfm of air at a minimum 90 psig pressure. Air is piped to the area from the air compressor through the following Schedule 40 pipe lengths and sizes: 200 ft of 2-in. pipe, 300 ft of $1\frac{1}{2}$-in. pipe, and 50 ft of $\frac{3}{4}$-in. pipe. Determine if the pipe sizing is adequate to prevent excessive pressure drop.

11-6.—A $1\frac{1}{4}$-in.-diameter tubing 25 ft in length has 400 scfm of air flowing through it. The upstream pressure is 200 psig. What is the downstream pressure?

chapter twelve

FLUID ENERGY, WORK, AND POWER

12-1 INTRODUCTION

The ability of fluid devices to do useful work economically and efficiently is responsible for their widespread use. Fluids are applied in different ways in these devices. In some devices the static pressure of the fluid is used to perform the desired work. In other types the velocity head of the fluid is used to do the work. Four important devices that use fluid energy are shown in Fig. 12-1.

In Chapter Five the common unit of mechanical work energy is defined as the foot pound. When work energy is related to time, the result is a unit of power. Power is thus defined as the time rate of doing work. For example, 250 ft lb of work done in 1 h is a unit of power. In Chapter Five other energy terms equivalent to mechanical work are mentioned. These are heat and electrical energy.

The basic power unit is the horsepower, defined as follows:

$$1 \text{ hp} = 550 \, \frac{\text{ft lb}}{\text{s}} = 33{,}000 \, \frac{\text{ft lb}}{\text{min}} \qquad (12\text{-}1)$$

The heat and electrical energy power equivalents are given as follows:

$$1 \text{ hp} = 746 \text{ W} = 42.4 \, \frac{\text{Btu}}{\text{min}} \qquad (12\text{-}2)$$

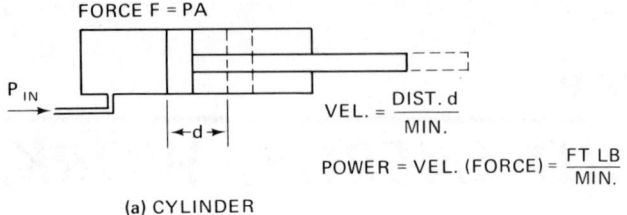
(a) CYLINDER

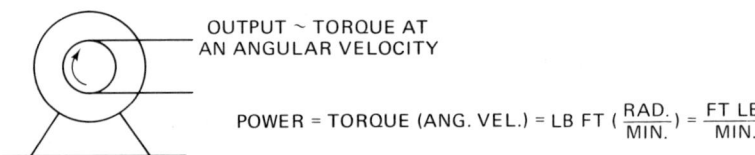
(b) PUMP

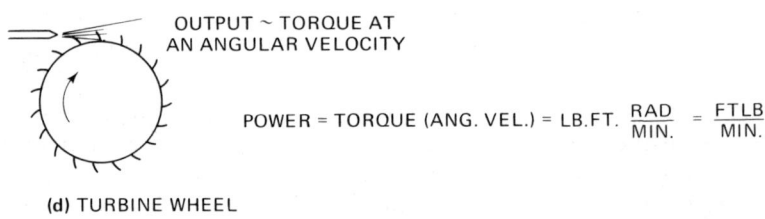
(c) HYDRAULIC OR PNEUMATIC MOTOR.

(d) TURBINE WHEEL

FIG. 12-1.

All the devices shown in Fig. 12-1 do mechanical work. Although the work output of the devices has different characteristics, it can be shown that all the outputs are reducible to the basic power term of foot pound per minute or per second. This is illustrated by the devices shown in Fig. 12-1.

12-2 PUMP HORSEPOWER

In Fig. 12-1(b), the work done by a pump is shown as the product of the net head, h, and the flow rate expressed in lb/min or lb/s. The horsepower

Sect. 12-2 / PUMP HORSEPOWER

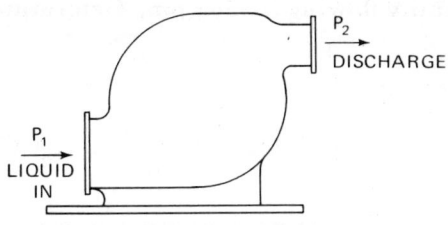

FIG. 12-2. Pump.

of the pump is determined by dividing into the pump output, the appropriate term from Eq. (12-1).

Although the cylinder, motor, and turbine shown in Fig. 12-1 may be operated by compressible or incompressible fluids, the pump works only with incompressible fluids. The air compressor might, for instance, be considered to do the equivalent job with a gas.

In Fig. 12-2, liquid enters the pump at the inlet at a certain pressure P_1. The pump does work on the liquid, increasing the pressure to P_2. The net head on the pump is $P_2 - P_1$. If P_1 is less than atmospheric and if gage pressures are being used, then P_1 is negative and the net head is the sum of the two, or $P_2 - (-P_1)$. If the net head is now expressed in terms of feet of the liquid, the total work done on the liquid is $h \times$ flow rate in lb/min or lb/s. The horsepower can then be put in the form of an equation:

$$\text{horsepower} = \frac{[(P_2/W_s) - (P_1/W_s)]G}{550 \text{ ft lb/s} - \text{hp}} \qquad (12\text{-}3)$$

where W_s = specific weight of the liquid
  G = flow rate, lb/sec

The term $(P_2/W_s) - (P_1/W_s)$ must be in feet in order to make the units consistent.

Note that the pressures P_1 and P_2 are those at the inlet and outlet at the pump and not at some distance away from the pump. The difference in elevation of the inlet and outlet is ordinarily so small that it may be neglected. Since the velocity head is reflected in the flow rate, all the requirements of Bernoulli's equation are met.

Equation (12-3) can be rewritten using the head, h, as follows:

$$\text{horsepower} = \frac{hG \text{ ft lb/s}}{550 \text{ ft lb/s} - \text{hp}} \qquad (12\text{-}4)$$

EXAMPLE

Oil with a specific gravity of 0.87 is being pumped through the pump in Fig. 12-2. The output pressure is 80 psig and the input pressure is

−3 psig. The quantity flowing is 300 gpm. Determine the horsepower output.

Solution

$$W_s = 0.87 \left(62.4 \, \frac{\text{lb}}{\text{ft}^3}\right) = 54.3 \, \frac{\text{lb}}{\text{ft}^3}$$

The 300-gpm flow rate must be converted to a weight basis, as follows:

$$G = \frac{(300 \text{ gpm})(54.3 \text{ lb/ft}^3)}{(7.48 \text{ gal/ft}^3)(60 \text{ s/min})}$$

$$= 36.3 \text{ lb/s}$$

The net head is

$$h = \frac{(80 \text{ lb/in.}^2)(144 \text{ in.}^2/\text{ft}^2)}{54.3 \text{ lb/ft}^3} - \frac{(-3 \text{ lb/in.}^2)(144 \text{ in.}^2/\text{ft}^2)}{54.3 \text{ lb/ft}^3}$$

$$= 212 \text{ ft} - (-8 \text{ ft})$$

$$= 220 \text{ ft}$$

The horsepower is

$$\text{horsepower} = \frac{(220 \text{ ft})(36.3 \text{ lb/s})}{550 \text{ ft lb/s-hp}} = 14.5 \text{ hp}$$

A simplified analogy which may be used to relate to pump power is to consider that the work done per unit time is equivalent to raising the liquid weight in lb/s a height equivalent to the net head h. This is illustrated in Fig. 12-3.

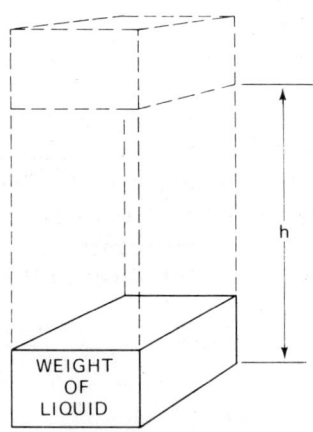

WT. OF LIQUID TRANSFERRED A HEIGHT h
IN A TIME INTERVAL ~ FT LB/MIN.

FIG. 12-3.

12-3 CYLINDER HORSEPOWER

The power output of a cylinder is the output force of the piston multiplied by the velocity with which it moves. This was illustrated in Fig. 12-1(a). The fluid pressure P exerted on the piston provides the output force. As described in Chapter Two, the force $F = PA$.

An example follows of a problem in which the solution depends on these principles.

EXAMPLE

The lift truck in Fig. 12-4 has a rated capacity of 2 tons. The platform is specified to be able to move up at a velocity of 1 ft/s with the 2-ton load in position. Determine (a) the horsepower and (b) the cylinder size if the pressure available is 150 psig.

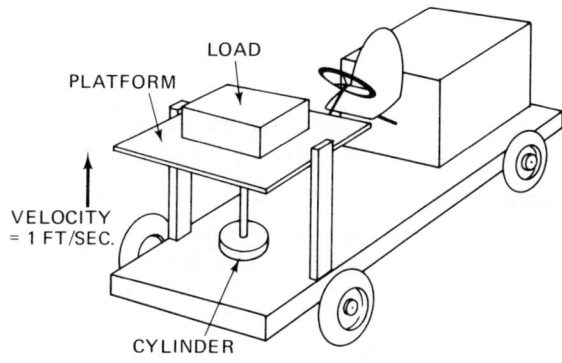

FIG. 12-4.

Solution

The output force is 2 tons, or 4000 lb. With the velocity of 1 ft/s, the time rate of doing work is (4000 lb)(1 ft)/s. The horsepower is

$$\text{horsepower} = \frac{4000 \text{ ft lb/s}}{550 \text{ ft lb/s}} = 7.27$$

The cylinder size is specified by the piston diameter. With $F = 4000$ lb and $P = 150$ psi, the area $A = 4000/150 = 26.7$ in². (The relationship $F = PA$ is used to determine this.) The diameter d is determined as follows:

$$A = \frac{\pi}{4} d^2 = 26.7$$

$$d = \sqrt{\frac{(4)(26.7)}{\pi}} = 5.83 \text{ in.}$$

A 6-in.-bore cylinder would be specified.

12-4 ENERGY INPUT AND BERNOULLI'S EQUATION

The pump power form giving work in time units, as ft lb/min, can be modified to an energy form that can be used in Bernoulli's equation. This now gives a convenient method to determine energy additions to a fluid system utilizing incompressible fluids. Restating Bernoulli's equation, it is

$$\begin{bmatrix}\text{press}\\\text{head}\end{bmatrix} + \begin{bmatrix}\text{vel.}\\\text{head}\end{bmatrix} + \begin{bmatrix}\text{elev.}\\\text{head}\end{bmatrix} + \begin{bmatrix}\text{energy}\\\text{added}\end{bmatrix}$$
$$= \begin{bmatrix}\text{press.}\\\text{head}\end{bmatrix} + \begin{bmatrix}\text{vel.}\\\text{head}\end{bmatrix} + \begin{bmatrix}\text{elev.}\\\text{head}\end{bmatrix} + \begin{bmatrix}\text{energy}\\\text{lost}\end{bmatrix}$$

All of these terms may be expressed in ft lb/lb of fluid flowing, as pointed out in Chapter Nine.

The pump power in ft lb/min can be converted to ft lb/lb of fluid by dividing the flow rate in lb/min into the power term. This is illustrated as follows:

$$\frac{\text{ft lb/min}}{\text{lb/min}} = \frac{\text{ft lb}}{\text{lb}}$$

This means that there is a method available for determining the power required for a pump to be used in a system where the other variables from Bernoulli's equation are known or can be estimated.

EXAMPLE

The system shown in Fig. 12-5 is set up to feed water into tank *A* at a pressure of 60 psig. With a constant flow rate of 200 gpm and with velocities, friction loss, and elevations as shown, what pump horsepower is required?

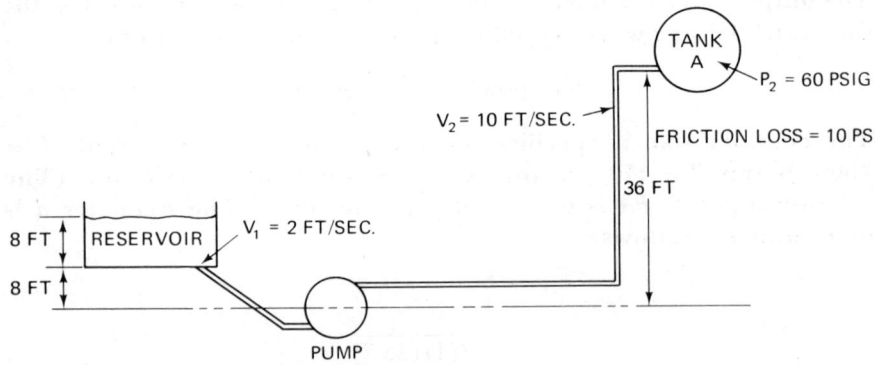

FIG. 12-5.

Sect. 12-5 / INPUT, OUTPUT AND EFFICIENCY

Solution

$$\frac{P_1}{W_s} + \frac{v_1^2}{2g} + Z_1 + h_p = \frac{P_2}{W_s} + \frac{v_2^2}{2g} + Z_2 + h_L \quad \bigg| (9\text{-}2)$$

Solving for h_p,

$$h_p = \frac{P_2}{W_s} - \frac{P_1}{W_s} + \frac{v_2^2}{2g} - \frac{v_1^2}{2g} + Z_2 - Z_1 + h_2$$

Substitution in the equation above yields

$$h_p = \frac{(60 \text{ lb/in.}^2)(144 \text{ in.}^2/\text{ft}^2)}{62.4 \text{ lb/ft}^3} - 8 \text{ ft} + \frac{(10 \text{ ft/s})^2}{(2)(32.2) \text{ ft/s}^2} - \frac{(2 \text{ ft/s})^2}{(2)(32.2) \text{ ft/s}^2}$$
$$+ 36 \text{ ft} - 8 \text{ ft} + \frac{(10 \text{ lb/in.}^2)(144 \text{ in.}^2/\text{ft}^2)}{62.4 \text{ lb/ft}^3}$$

Note that the P_1/W_s term is the 8 ft of water head on the pump.

$$h_p = 183 \text{ ft lb/lb of fluid flowing}$$

The pump should add this amount of energy to each pound of water flowing in the system. It now remains to determine the number of pounds flowing per second.

$$G = \frac{(200 \text{ gal/min})(62.4 \text{ lb/ft}^3)}{(7.48 \text{ gal/ft}^3)(60 \text{ s/min})} = 27.8 \frac{\text{lb}}{\text{s}}$$

From Eq. (12-3),

$$\text{horsepower} = \frac{(183 \text{ ft lb/lb})(27.8 \text{ lb/s})}{550 \text{ ft lb/s}}$$
$$= 9.25$$

In practice a 10-hp pump would be selected.

12-5 INPUT, OUTPUT, AND EFFICIENCY

The efficiency of a machine is important to its economic utilization. We can define efficiency by stating that it is a comparison of the power output to the power input, as shown in Fig. 12-6. The power is put into the pump at the shaft and is taken out at the discharge of the pump. When expressed in common units, a comparison can be made as shown in the next paragraph.

With this in mind, we can make a more precise definition of effi-

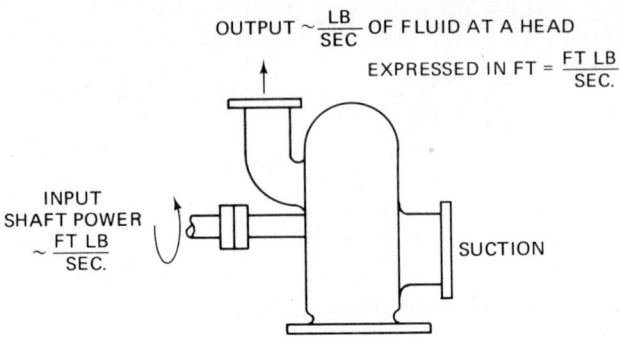

FIG. 12-6. Efficiency terms for a pump.

ciency. Efficiency is expressed in percentage and is the output divided by the input multiplied by 100 percent:

$$\text{efficiency} = \frac{\text{output}}{\text{input}} \times 100\% \tag{12-5}$$

As long as the units of input and output are the same, any power term may be used.

EXAMPLE

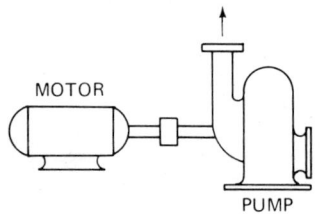

FIG. 12-7.

The electric motor driving the pump in Fig. 12-7 requires 8000 W of power. The motor efficiency is 95 percent. The pump output is 8 hp. What is the efficiency of the pump?

Solution

The efficiencies of the motor and the pump are both involved. The power input into the pump is the motor output. This is 8000 W × motor efficiency, or (8000)(0.95) = 7600 W. Converting this horsepower,

$$\text{pump input power} = \frac{7600 \text{ W}}{746 \text{ W/hp}} = 10.2 \text{ hp}$$

$$\text{pump efficiency} = \frac{(8)(100\%)}{10.2} = 78.4\%$$

QUESTIONS AND PROBLEMS

12-1.—An electric motor is rated at 9540 W power output. What is its horsepower?

12-2.—The pump in Fig. 12-8 pumps water at the rate of 250 gpm. The pressure on the discharge end is 35 psig. What is the pump horsepower?

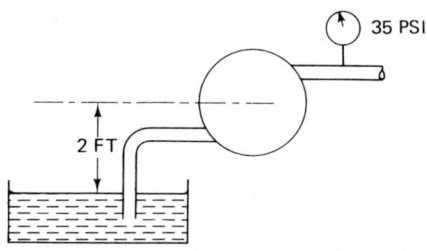

FIG. 12-8.

12-3.—The pump shown in Fig. 12-9 has an output of 250 gpm of water. Its efficiency is 52 percent. What is the output pressure?

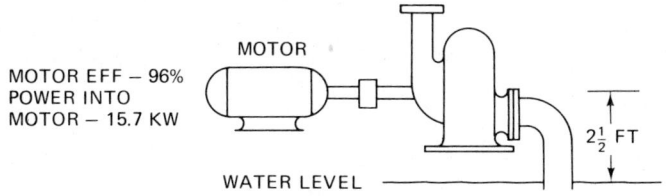

FIG. 12-9.

12-4.—The piston in the cylinder in Fig. 12-10 moves at the rate of 6 ft/min against a load resistance of 1230 lb. How much work is done in $\frac{1}{2}$ min?

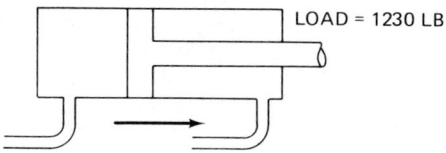

FIG. 12-10.

12-5.—The pump in Fig. 12-11 boosts the pressure of hydraulic oil (specific gravity of 0.87) as shown. The pump efficiency is 55

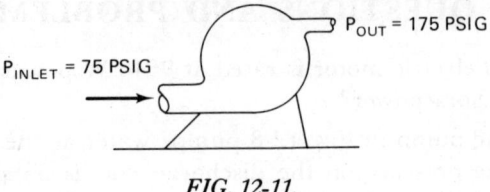

FIG. 12-11.

percent. If it requires 2.5 hp to drive the pump, what is the flow rate in gpm?

12-6.—A pump puts 2 hp into the fluid in the system in which it is installed. Forty Btu/min is lost due to heat caused by friction of the fluid plus mechanical friction. What is the horsepower output of the pump?

chapter thirteen
IMPULSE AND MOMENTUM

13-1 INTRODUCTION

The impulse–momentum principle is made use of in fluid mechanics for many applications. Two applications of it are the gas turbine used for aircraft and the steam turbine used in electric power generation. To understand this principle it is necessary to return to physics for a definition.

The *momentum* of an object is defined as the product of its mass and its velocity. Momentum is a vector quantity because velocity is a vector quantity. A *vector* is a quantity that has both magnitude and direction. For example, a velocity of 10 ft/s is not described completely until the direction is known. If the 10 ft/s velocity is in a vertical direction and upward, the description is complete. If the direction or the magnitude of the velocity is changed, a change in the vector quantity describing the velocity results.

Vectors are represented graphically by straight lines. The length of the line is proportional to the magnitude. The direction is indicated by the angle the line makes with a reference surface and by an arrowhead. The 10 ft/s upward velocity is indicated vectorially in Fig. 13-1(a). Examples of other vectors are shown in Fig. 13-1(b) and (c).

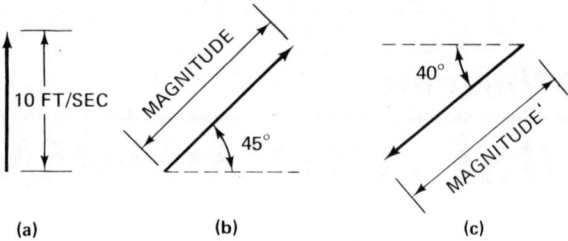

FIG. 13-1. Vectors.

Vector components having the same direction can be algebraically added and subtracted. Examples of vector addition are shown in Fig. 13-2. In Fig. 13-2(a), vectors representing the quantities of 5 and are shown. They are in the same direction; therefore, the sum of the two is 5 + 5 = 10.

In Fig. 13-2(b), the component of B which is in the same direction as A must be determined if vector addition involving A is done. This component is $B \cos \theta$. The sum of the two is then $A + B \cos \theta$. The addition of the two vectors C and D in Fig. 13-2(c), can be done by considering that vector D has a negative sign, since its direction is 180° from C. The sum of the two is then $10 + (-4) = 6$.

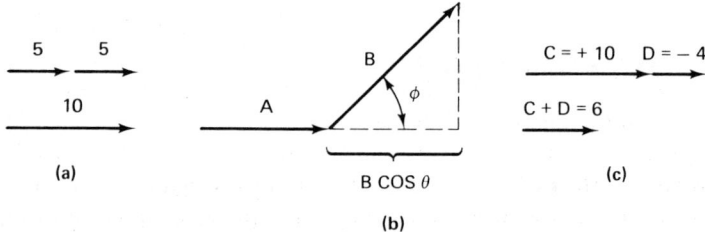

FIG. 13-2. Vector addition.

The impulse of a force is defined as the product of a resultant force and a time interval or increment Δt (delta t). When an impulse is applied to a body having momentum, the momentum of the body is changed. The relationship between the impulse and the momentum is as follows: change in momentum of a body = impulse of the force acting on it. Rearranging this and expressing it mathematically:

$$F \Delta t = \Delta(Mv) \tag{13-1}$$

Δt and ΔMv represent time and momentum changes, respectively. Since the mass of the body cannot be changed, the right-hand term of the equation can be changed to $M \Delta v$. The equation can be rewritten as

Sect. 13-2 / IMPULSE TURBINE

$$F \Delta t = M \Delta v \tag{13-2}$$

In Fig. 13-3 three examples of the impulse–momentum principle are shown. The bat strikes the baseball with a force F and during the time increment Δt the impulsive force causes a change in velocity of the ball and a change in momentum.

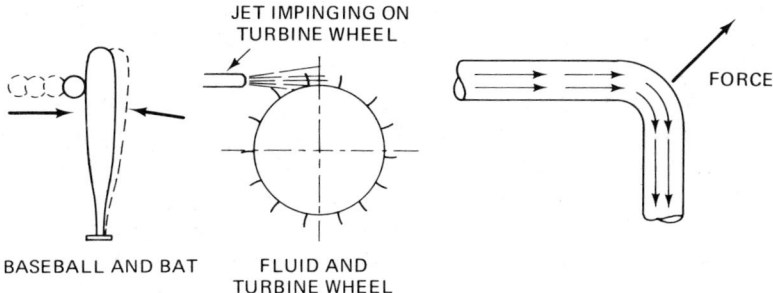

FIG. 13-3. Examples illustrating the impulse-momentum principle.

The other two examples illustrate the change of momentum that occurs in fluid in motion. The turbine wheel uses the change in momentum of the fluid to do useful work. The change in momentum of the fluid in the pipe bend causes a resultant force F which stresses the pipe.

There are two areas in which the impulse–momentum principle is of concern. One is the practical application of it for the turbine. The second is the consideration of forces and stresses resulting from the change in momentum when fluids change velocity.

13-2 IMPULSE TURBINE

The *impulse turbine principle* is applied from steam-turbine power-plant turbines all the way down in size to small air-driven turbines used in hand tools. A high percentage of gyroscopes made for aircraft instruments are powered by a jet generated by vacuum system.

The force exerted on a turbine blade by a jet can be developed by first

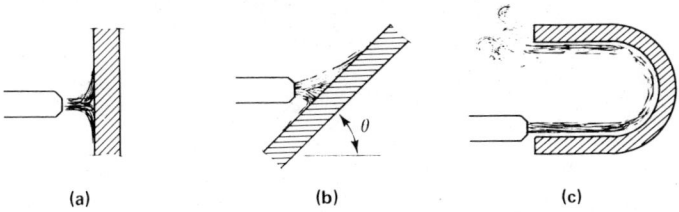

FIG. 13-4. Jet impingement on different stationary surfaces.

considering the impingement of a jet on stationary surfaces of various forms. In Fig. 13-4 are shown three jets striking three plates. The plates in Fig. 13-4(a) and (b) are both flat but have different angles with respect to the jet. The third, part (c), is shaped so that the flow of the jet is reversed.

Assuming a constant mass rate of flow of the same amount in all three cases, the difference in impulse of the three can be caused by differences in the velocity changes. Now consider what these velocity changes are. In Fig. 13-5 the velocity changes that occur in these three jets are shown. The initial jet velocity is v_0 and the velocity of the jet after striking the plate or vane is notated v_1 and v_2. In all cases it is assumed that there is no friction loss as the jet flows along the surface of the plate or vane. Therefore, the *magnitude* of the velocities equals that of the initial velocity.

The force acting on the three plates is also shown in Fig. 13-5. This force is perpendicular to the flat surfaces in (a) and (b). The vane in (c)

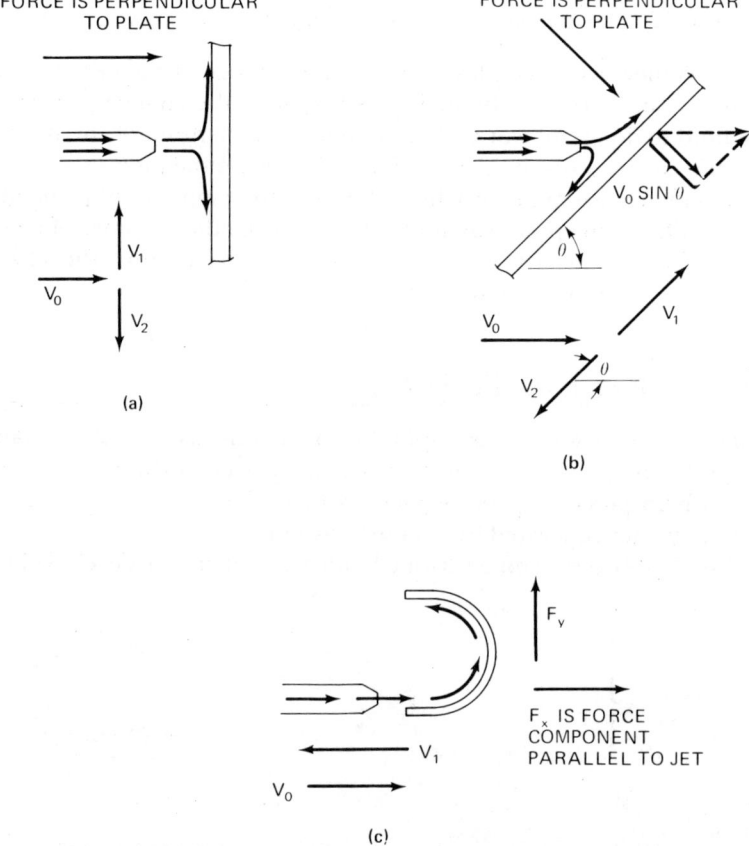

FIG. 13-5. Jet impingement on stationary surfaces.

Sect. 13-2 / IMPULSE TURBINE

changes the direction of the jet and redirects it 180° from its original direction.

The analysis of the force exerted by a jet then is based on consideration of the following:

1 / **The change in velocity Δv is the vector difference of the initial velocity and the final velocity.**

2 / **The velocity terms are the vector components parallel to the force.**

These statements can be explained further by considering each of the three jets in Fig. 13-5. The explanations of the three jets are accompanied by additional sketches designated as Fig. 13-5 (a'), (b'), and (c').

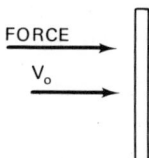

FIG. 13-5a'.

The final velocity component in the direction of the force is zero. Then

$$\Delta v = v_0 - 0 \tag{13-3}$$
$$\Delta v = v_0$$

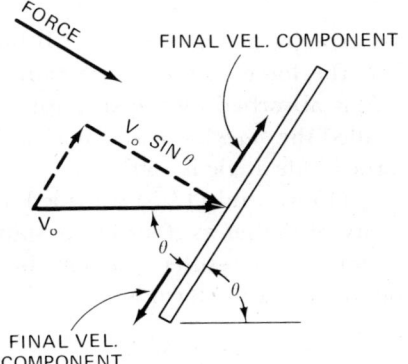

FIG. 13-5b'.

$$\Delta v = v_0 \sin \theta - 0 \tag{13-4}$$
$$\Delta v = v_0 \sin \theta$$

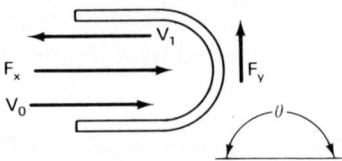

FIG. 13-5c'.

$$\Delta v = v_0 = (-v_1) \qquad (13\text{-}5)$$
$$= v_0 + v_1$$

With no losses, $v_0 - v_1$

Then $\Delta v = 2v_0$ with the vane stationary.

If $\theta \neq 180°$ then
$$\Delta v = v_0 - v_1 \cos\theta$$

But $v_0 = v_1$ and
$$\Delta v = v_0 - v_0 \cos\theta = v_0(1 - \cos\theta)$$

The jet impinges on the stationary plate in Fig. 13-5 (a′) at an angle of 90°. The velocity change is 90° from the original vector. The final velocity has no component in the direction of the force and is therefore zero.

The jet in Fig. 13-5(b′) impinges on the flat plate at an angle θ. The resultant force on the plate is at an angle of 90°. The initial velocity component in this direction is $v_0 \sin\theta$. As in the case of the jet striking the plate at 90°, the final velocity components are parallel to the plate and equal to zero.

The vane reverses the jet in Fig. 13-5(c′). If the vane is a part of a turbine wheel or rotor, the force in the x direction is that which accomplishes useful work. F_y is absorbed by the structure and is not considered here. The angle θ is called the *blade angle* and here is 180°. The force F_x is at a maximum when the blade angle is 180°.

Equations (13-3), (13-4), and (13-5) provided a means to determine the difference in velocity of the jet as it strikes a stationary surface of the type described. The force that results can now be developed from Eq. (13-2). This equation in its original form is

$$F \Delta t = M \Delta v$$

Rearranging,

$$F = \frac{M \Delta v}{\Delta t} \qquad (13\text{-}6)$$

But $M/\Delta t$ is the mass flow rate and by using W/g in place of M, the weight flow rate in lb/s can be used in the equation. With G representing weight flow rate, Eq. (13-6) can then be changed to

$$F = \frac{G \Delta v}{g} \qquad (13\text{-}7)$$

Using this form, the equations for the force exerted on each of the surfaces in Fig. 13-5 are as follows:

(a) Jet at 90° angle:

$$F = \frac{G v_0}{g} \qquad (13\text{-}8)$$

(b) Jet at angle θ to a flat plane:

$$F = \frac{G v_0 \sin \theta}{g} \qquad (13\text{-}9)$$

(c) Stationary vane with angle θ:

$$F_x = \frac{G v_0 (1 - \cos \theta)}{g} \qquad (13\text{-}10)$$

F_x is the force component in the x direction.

13-3 JET IMPINGEMENT ON MOVING SURFACES

The analysis of jet velocities in Fig. 13-5 is based on the assumption of stationary surfaces. In any fluid machine, such as a turbine, this is not true and the turbine rotates. The net force on the surface then depends on the *relative* velocity of the jet with respect to the surface, as shown in Fig. 13-6.

With v_0 as the jet velocity and v_a as the vane velocity, the relative velocity of the jet with respect to the vane is $v_0 - v_a$. Letting u equal the relative velocity, $u = v_0 - v_a$. The initial velocity with which the jet strikes the vane then is the relative velocity u, as indicated in Fig. 13-6(a). The relative velocity here is represented by a vector. The velocity of the jet at any point after it strikes the vane is the vector sum of the relative velocity and the vane velocity. A graphical representation of this vector addition is shown in Fig. 13-6(b).

Considering only the force in the x direction, the velocity change Δv can be determined in a manner similar to that for the stationary vane. As shown in Fig. 13-6(c), the initial velocity is the relative velocity, $v_0 - v_a$. The final velocity is the component of the relative velocity in the x direction, $(v_0 - v_a) \cos \theta$.

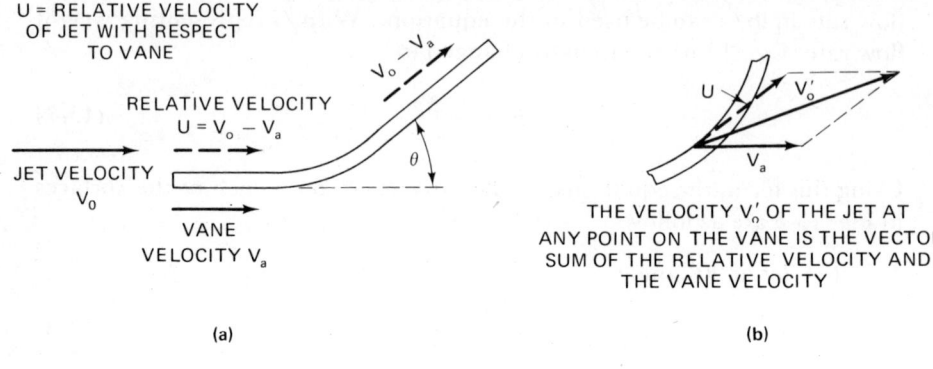
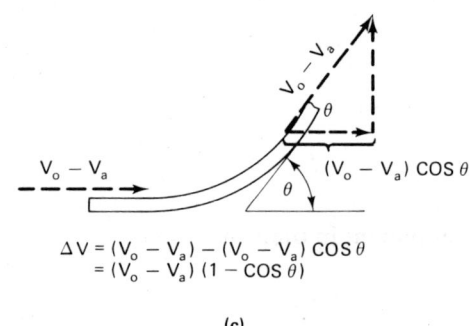

FIG. 13-6. Jet velocity and a moving vane.

The equation then becomes

$$\Delta v = (v_0 - v_a)(1 - \cos \theta) \tag{13-11}$$

The force on the vane is

$$F_x = \frac{G(v_0 - v_a)(1 - \cos \theta)}{g} \tag{13-12}$$

13-4 WORK DONE ON A MOVING SURFACE

The basic work relationship of work being equal to a force multiplied by a distance can be used to develop an equation for the work done on a moving vane. The only force component doing work is F_x, since F_y is absorbed by the rotor and the bearings. The distance is the linear velocity v_a of the vane. The equation for the work is

Sect. 13-5 / FORCES ON PIPE BEND

$$\text{work} = F_x v_a$$

$$= \frac{G(v_0 - v_a)(1 - \cos\theta)v_a}{g} \qquad (13\text{-}13)$$

The distance portion of this relationship is based on distance per unit time, or velocity. Therefore, the work done is work per unit time. The horsepower can then be determined by dividing by the work equivalent of 1 hp. With the units of feet and seconds, this gives

$$\text{horsepower} = \frac{G(v_0 - v_a)(1 - \cos\theta)v_a}{550g} \qquad (13\text{-}14)$$

13-5 FORCES ON PIPE BEND WITH ENLARGING OR CONTRACTING SECTIONS

The impulse–momentum principle can be used to compute forces on pipe sections where bends, enlargements, or contractions occur. The forces that occur with pressurized fluids flowing in a pipe bend are the result of the internal pressure plus that caused by the change in momentum as the fluid changes velocity. This is more important where liquids are concerned because of the greater mass as compared to gases.

Figure 13-7(a), shows a pipe bend with a reducing diameter with a fluid entering it at an initial velocity v_1 and a final velocity v_2. The magnitude of the velocity changes because of the reduced pipe section at the cross-sectional area A_2. The analysis of the forces is made using the principles of Eq. (13-7) and the force diagram of Fig. 13-7(b). In Eq. (13-7), the force F is the resultant of all the forces acting in any given direction. The equation can then be written as follows:

$$\Sigma F = \frac{G \Delta v}{g} \qquad (13\text{-}7)$$

The forces in the x and the y directions are used with Eq. (13-7) in setting up the solution.

The forces acting on the fluid in the bend are shown in Fig. 13-7(b). F_x and F_y are the components of the force exerted by the pipe bend on the fluid. In Fig. 13-7(c), the vectors and resulting algebraic expressions are shown. The signs of F_x and F_y are assumed positive or negative as shown. If the assumption is wrong, the sign of the solved terms will be negative. The force summation in each direction is then equated to the right-hand term of Eq. (13-7). The velocity changes in this term are the vector components in the x and y directions. Putting these in the final form, the equations are as follows:

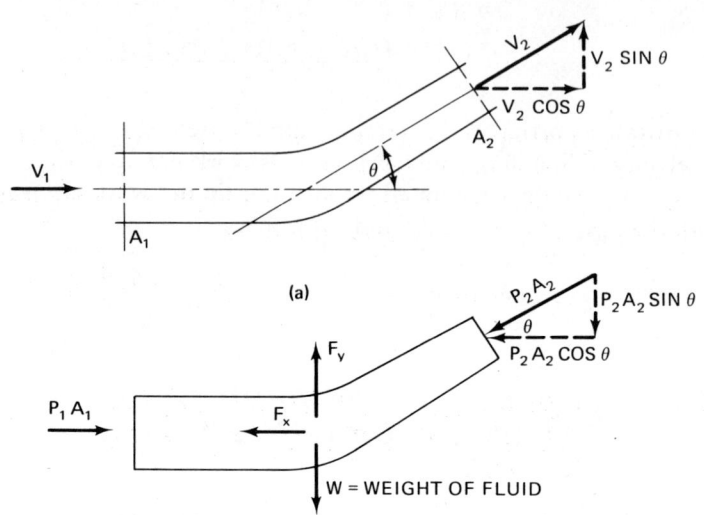

(a)

(b) FORCES ACTING ON FLUID IN BEND

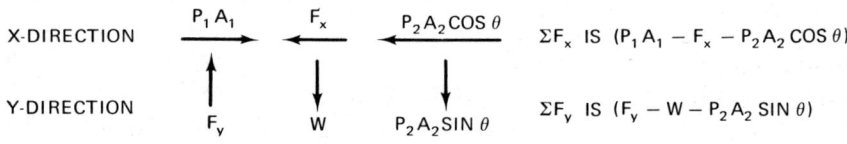

(c) X- AND Y- FORCE COMPONENTS

FIG. 13-7.

In the x direction:

$$P_1 A_1 - F_x - P_2 A_2 \cos\theta = \frac{G}{g}(v_2 \cos\theta - v_1) \qquad (13\text{-}15)$$

In the y direction:

$$F_y - W - P_2 A_2 \sin\theta = \frac{G}{g}(v_2 \sin\theta - 0) \qquad (13\text{-}16)$$

EXAMPLE 1

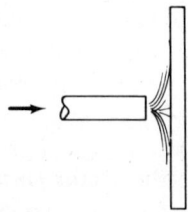

Water with a velocity of 20 ft/sec strikes the flat plate in Fig. 13-8. The cross-sectional area of the pipe is 2.3 in.^{2}. Determine the force exerted on the plate.

FIG. 13-8.

Sect. 13-5 / FORCES ON PIPE BEND

Solution

This is the same condition as shown in Fig. 13-5(a), and Eq. (13-8) applies. The equation is

$$F = \frac{Gv_0}{g}$$

The equation of continuity is used to determine G, the weight of the flow. The volume flow rate, Q, when multiplied by the specific weight, W_s, yields the rate of flow in lb/s:

$$Q = Av_0 = \left(\frac{2.3 \text{ in.}^2}{144 \text{ in.}^2/\text{ft}^2}\right)\left(20\frac{\text{ft}}{\text{s}}\right)$$

$$= 0.32 \frac{\text{ft}^3}{\text{s}}$$

$$G = \left(0.32\frac{\text{ft}^3}{\text{s}}\right)\left(62.4\frac{\text{lb}}{\text{ft}^3}\right) = 19.9\frac{\text{lb}}{\text{s}}$$

and

$$F = \frac{\left(19.9\frac{\text{lb}}{\text{s}}\right)\left(20\frac{\text{ft}}{\text{s}}\right)}{32.2 \text{ ft/s}^2} = 12.4 \text{ lb}$$

EXAMPLE 2

Determine the force on the moving vane in Fig. 13-9 with a liquid of 0.86 specific gravity flowing through the nozzle. The area of the pipe is 0.012 ft².

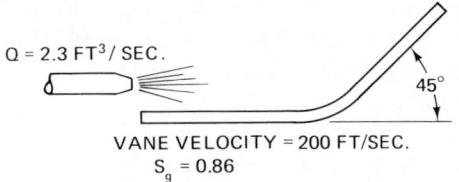

FIG. 13-9.

Solution

Equation (13-12) applies, since the force in the x direction is desired. This equation is

$$F = \frac{G(v_0 - v_a)(1 - \cos\theta)}{g}$$

The vane velocity v_s is 200 ft/s. The velocity of the jet, v_0, is not known but can be determined from the equation of continuity, $Q = Av$.

$$v_0 = \frac{Q}{A} = \frac{2.3 \text{ ft}^3/\text{s}}{0.012 \text{ ft}^2} = 192 \frac{\text{ft}}{\text{s}}$$

This velocity is less than the velocity of the vane. Consequently, the jet can exert no force on the vane and some other force must be driving the vane at 200 ft/s.

EXAMPLE 3

Determine the forces in the horizontal and vertical planes which act on the pipe bend in Fig. 13-10. The volume of the bend is 0.9 ft³. The fluid is water.

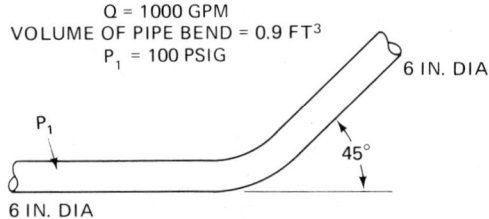

FIG. 13-10.

Solution

The forces acting on the water volume are shown in Fig. 13-7(b). Note that the diameter is constant through the pipe. This means that v_1 equals v_2, since $Q = A_1 v_1 = A_2 v_2$ and $A_1 = A_2$. The pressure difference resulting from elevation differences is so small that it may be neglected. Therefore, $P_1 = P_2$. Equations (13-15) and (13-16) apply. Equation (13-15) reads

$$P_1 A_1 - F_x - P_2 A_2 \cos\theta = \frac{G}{g}(v_1 \cos\theta - v_2)$$

$P_1 = 100 \text{ psi} \qquad A_1 = 28.3 \text{ in.}^2 = \frac{28.3}{144} \text{ ft}^2 = 0.197 \text{ ft}^2$

$Q = \frac{1000}{(7.48)(60)} = 2.22 \frac{\text{ft}^3}{\text{s}} \qquad G = (2.22)(62.4) = 138 \frac{\text{lb}}{\text{s}}$

$v_1 = \frac{Q}{A_1} = \frac{2.22}{0.197} = 11.3 \frac{\text{ft}}{\text{s}}$

QUESTIONS AND PROBLEMS

Substitution yields

$$(100)(28.3) - F_x - (100)(28.3)(0.707)$$
$$= \frac{138}{32.2}[(11.3)(0.707) - 11.3]$$
$$F_x = 843 \text{ lb}$$

The sign of F_x is positive, so the assumed direction is correct. Equation (13-16) is used for the y direction.

$$F_y - W - P_2 A_2 \sin\theta = \frac{G}{g}(v_2 \sin\theta - 0)$$
$$W = (0.9)(62.4) = 56.1 \text{ lb}$$

Substitution yields

$$F_y - 56.1 - (100)(28.3)(0.707) = \frac{138}{32.2}(11.3)(0.707)$$
$$F_y = 2091 \text{ lb}$$

The sign is positive, so the direction is correct. The forces that act on the pipe bend are *equal* and *opposite* to the direction shown for F_x and F_y. This is true since the forces solved for are those which act on the water.

QUESTIONS AND PROBLEMS

13-1.— A jet of water strikes the flat plate as shown in Fig. 13-11. With the quantity and area indicated, calculate the force with which it hits the plate.

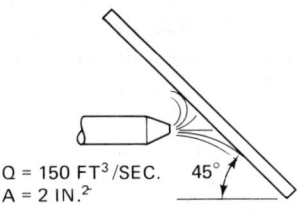

Q = 150 FT³/SEC. 45°
A = 2 IN.²

FIG. 13-11.

13-2.—

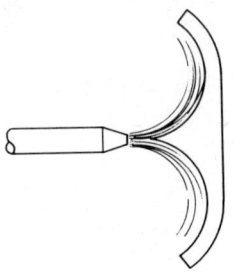

FIG. 13-12.

The split vane in Fig. 13-12 is used in impulse turbine wheels. What is the resultant force in the y direction that the jet exerts on the vane? Why would this type of construction be of advantage?

13-3.—

JET VEL. = 170 FT/SEC.
G = 50 LB/SEC.

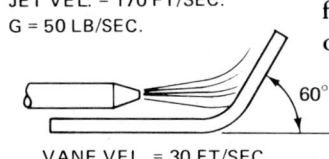

VANE VEL. = 30 FT/SEC.

FIG. 13-13.

A water jet impinges on a series of turbine vanes as shown in Fig. 13-13. The flow rate is 50 lb/s. Calculate the force on the vanes and the horsepower.

13-4.—

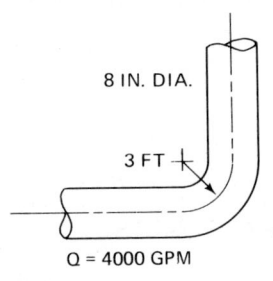

8 IN. DIA.

3 FT

Q = 4000 GPM

FIG. 13-14.

The 8-in.-diameter pipe in Fig. 13-14 has a 90° bend as shown and carries 4000 gpm of water. The pressure is 100 psig. Calculate the x and y components of the force on the pipe. (*Note:* Calculate the volume of the bend using the cross-sectional area and a mean length equal to the circumference of the 3-ft bend.)

13-5.—

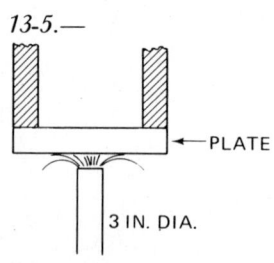

← PLATE

3 IN. DIA.

FIG. 13-15.

What velocity and flow rate of water are required to hold the flat plate in Fig. 13-15 against the surfaces shown? The plate weighs 50 lb.

chapter fourteen

ORIFICES, VENTURIS, AND PITOT TUBES

14-1 INTRODUCTION

A common definition of *orifice* is simply "opening" or "hole." However, in engineering practice, the orifice is a specific device having designed characteristics. It usually is designed into a flat plate and consists of a circular hole in this plate. Its most common use is as a flow quantity measuring device. An orifice may, as shown in Fig. 14-1, be installed in a pipeline to measure the flow rate. The rounded-edged orifice shown in Fig. 14-2 is another type.

The determination of flow quantity using an orifice is based on the

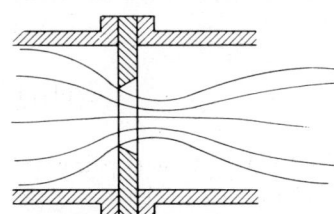

FIG. 14-1. Sharp-edged orifice.

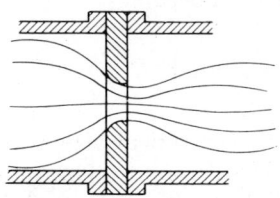

FIG. 14-2. Round-edged orifice.

use of Torricelli's theorem and also Bernoulli's equation. From Chapter Eight, the equation expressing Torricelli's theorem is

$$v = \sqrt{2gh} \tag{8-6}$$

where v is the theoretical velocity of a free jet of fluid discharging into the atmosphere under a head h. With the proper modifications, this theorem may be used to measure the fluid velocity flowing through an orifice. The equation of continuity, $Q = Av$, is the basis for determining the flow quantity.

14-2   ORIFICE FLOW CHARACTERISTICS

The actual flow characteristics of fluid through an orifice vary from the theoretical as shown in Fig. 14-3. The fluid flows through the orifice and then continues to reduce in size until a minimum size is reached at a point called the *vena contracta*. The cross-sectional area of the jet of fluid here is designated A_2. The area of the orifice is designated A and that upstream from the orifice is A_1. The diameter of the orifice is d. Experiments have shown that the vena contracta is located a distance $d/2$ downstream from the orifice.

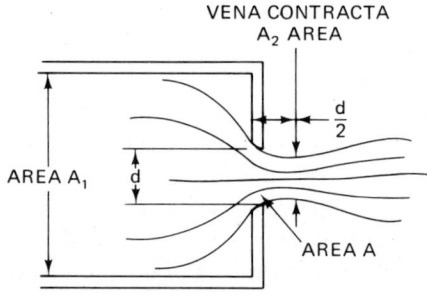

FIG. 14-3.   Orifice flow.

The coefficient of contraction is defined as the ratio of the areas of the jet, A_2, to the orifice, A. Using C_c to indicate the coefficient of contraction, the equation becomes

$$C_c = \frac{A_2}{A} \tag{14-1}$$

The theoretical velocity of the jet is equal to $\sqrt{2gh}$, according to Torricelli's theorem. The actual velocity is lower than this. Using v_a as the actual velocity and v as the theoretical velocity, the coefficient of velocity C_v is

Sect. 14-2 / ORIFICE FLOW CHARACTERISTICS

$$C_v = \frac{v_a}{v} \qquad (14\text{-}2)$$

The theoretical quantity of flow through the orifice is equal to the orifice area, A, multiplied by the theoretical velocity v. The coefficients of contraction and velocity may now be used to modify the theoretical equation to fit actual conditions:

$$Q = C_c A C_v \sqrt{2gh} \qquad (14\text{-}3)$$

The coefficients may be combined into a third coefficient called simply the *coefficient of discharge*, C_D. This is expressed by

$$C_D = C_c C_v \qquad (14\text{-}4)$$

Equation (14-3) may now be simplified to give the following:

$$Q = C_D A \sqrt{2gh} \qquad (14\text{-}5)$$

In practice, the coefficient of discharge has to be determined empirically for each installation. The coefficient of contraction, C_c, depends on the ratio of orifice area A to the upstream pipe area, A_1. For ratios in common use, C_c may vary from 0.60 to approximately 0.65.

The velocity coefficient, C_v, is closer to 1 and may range from 0.95 to 0.99. The range of the coefficients of discharge, C_D, then, are approximately 0.57 to 0.64.

Equations (14-3) and (14-4) apply for the case of the jet discharging to atmosphere. If an orifice is used in a closed pipeline such as shown in Fig. 14-4, the velocity of the fluid has an effect on the flow quantity through the orifice. Equations (14-4) and (14-5) then cannot be used in their present form.

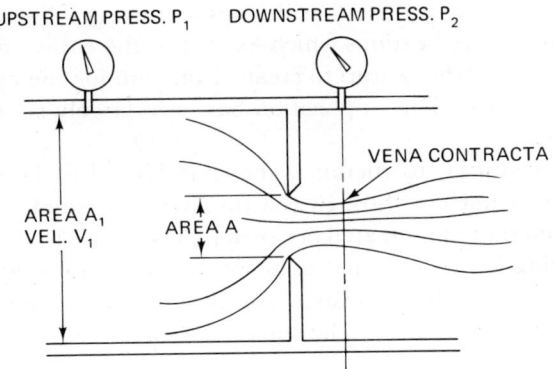

FIG. 14-4. Orifice in a pipeline.

The equations can be adapted for use by considering that the head, h, is made up of the pressure differences, $P_1 - P_2$, plus the velocity head at the upstream point. Using Eq. (14-5), the term for h then is

$$\left(\frac{P_1}{W_s} - \frac{P_2}{W_s}\right) + \frac{v_1^2}{2g}$$

Substituting this, the equation then becomes

$$Q = C_D A \sqrt{2g\left[\left(\frac{P_1}{W_s} - \frac{P_2}{W_s}\right) + \frac{v_1^2}{2g}\right]}$$

The flow Q upstream also equals $A_1 v_1$, or $Q = A_1 v_1$. Solving for v_1, $v_1 = Q/A_1$. The term Q/A_1 is now substituted for v_1 in the preceding equation. When this is solved for Q, the final equation is

$$Q = \frac{C_D A}{\sqrt{1 - (A/A_1)^2}} \sqrt{2g\left(\frac{P_1}{W_s} - \frac{P_2}{W_s}\right)} \qquad (14\text{-}6)$$

To simplify the final form of the equation, C_D is dropped before the substitution is made and then replaced after the solution is complete. Equation (14-6) is used for incompressible fluid flow only.

Orifices have the disadvantage of a relatively large pressure drop which occurs across the orifice. Other flow-measuring devices, such as the venturi, offer better characteristics in this respect.

14-3 VENTURI TUBES

The *venturi* tube has other uses besides being used as a flowmeter. One of the most common applications which existed in the earlier days of the airplane was the use of the venturi to create a vacuum for the operation of the aircraft instruments. This application has been largely supplanted by the vacuum pump.

Typical venturi construction is shown in Fig. 14-5. Fluid flow is from the large-angled portion through the throat to the small-angled portion. A pressure differential is created between points 1 and 2, with the pressure at point 2 being lower than that at point 1. Bernoulli's equation can be used to prove this, as shown below. The elevation terms are omitted, since the installation is horizontal. The equation then is

$$\frac{P_1}{W_s} + \frac{v_1^2}{2g} = \frac{P_2}{W_s} + \frac{v_2^2}{2g}$$

Sect. 14-4 / PITOT TUBES

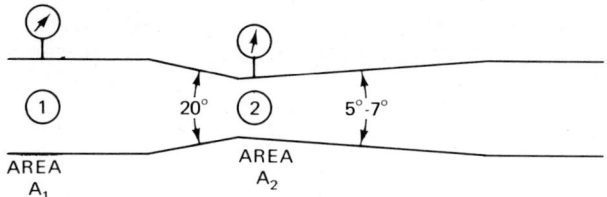

FIG. 14-5. Venturi.

Transposing,

$$\frac{P_1}{W_s} - \frac{P_2}{W_s} = \frac{v_2^2}{2g} - \frac{v_1^2}{2g}$$

Inspection of this equation confirms the pressure differential, since the velocity at point 2 has to be greater than the velocity at point 1.

The development of the equation for the flow quantity of an incompressible fluid in a venturi is similar to that for an orifice. The form of the equation is also similar to that for the orifice:

$$Q = \frac{CA_2}{\sqrt{1 - (A_2/A_1)^2}} \sqrt{2g\left(\frac{P_1}{W_s} - \frac{P_2}{W_s}\right)} \qquad (14\text{-}7)$$

The coefficient C for the venturi has typical values ranging from 0.89 to 0.99, depending on the Reynolds number of the flowing fluid.

Venturi meters have the advantage of a smaller pressure drop across the meter than orifices. In practice the pressure taps at point 1 and 2 in Fig. 14-5 may be individual gages or a differential manometer may be connected between the two points.

14-4 PITOT TUBES

The pitot-static tube, as shown in Fig. 14-6, is designed for velocity measurement instead of flow quantity. Pitot-static tubes are used for measurement of the flow of both compressible and incompressible fluids. The airspeed indicator used on aircraft is an example of an application of the pitot-static tube.

The equation for the velocity of incompressible flow in a pitot-static tube is a variation of the form of Torricelli's equation, $v = \sqrt{2_g h}$. Using the nomenclature of Fig. 14-6, the equation is

$$v = \sqrt{2_g \left(\frac{P_T}{W} - \frac{P_S}{W}\right)} \qquad (14\text{-}8)$$

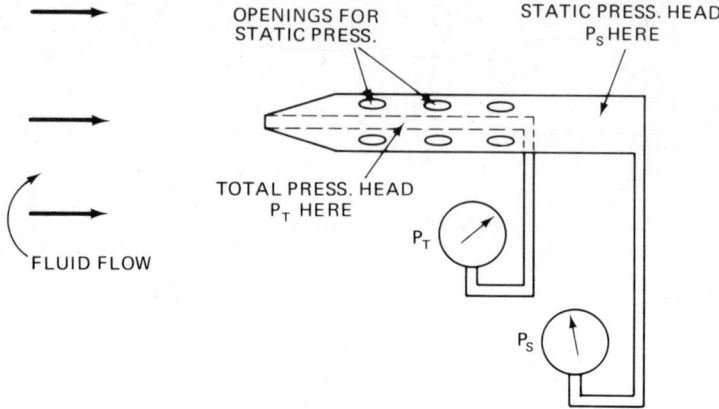

FIG. 14-6. Pitot-static tube.

Referring again to Fig. 14-6, the static pressure of the fluid at some distance upstream from the pitot-static tube is P_S. The sum of the static pressure and the velocity head is represented by P_T.

In Fig. 14-6, the inside tube is open at the small end of the left. The moving fluid impacts on the open end, providing a means to measure the total pressure, P_T. The larger-diameter tube is sealed at the left, so the velocity head has no effect. The static opening holes are at right angles to the fluid flow. In this position the only pressure sensed is that caused by the static pressure P_S. Individual gages or a differential manometer may be used to measure the pressures P_T and P_S.

EXAMPLE 1

The orifice in Fig. 14-7 is installed in a pipe with areas as shown. With water flowing, the orifice discharge was checked experimentally and found to be 5 ft³/s. Determine the coefficient of discharge.

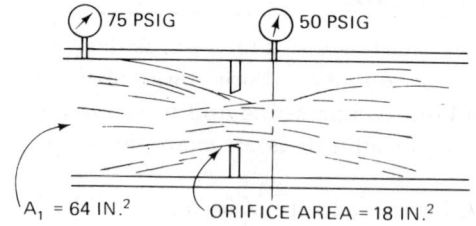

FIG. 14-7.

Solution

This orifice is located in a pipeline, so Eq. (14-6) applies. This is shown again.

Sect. 14-4 / PITOT TUBES

$$Q = \frac{C_D A}{\sqrt{1 - (A/A_1)^2}} \sqrt{2g \left(\frac{P_1}{W_s} - \frac{P_2}{W_s} \right)}$$

The actual discharge, Q, has already been determined by experiment. The problem is to find the value of C_D. The theoretical discharge may be determined by using the equation without C_D in it. The coefficient of discharge, C_D, then, is the actual discharge divided by the theoretical discharge. The theoretical discharge is

$$Q_{theor} = \frac{A}{\sqrt{1 - (A/A_1)^2}} \sqrt{2g \left(\frac{P_1}{W_s} - \frac{P_2}{W_s} \right)}$$

Substituting,

$$Q_{theor} = \frac{\frac{18}{144} \text{ft}^2}{\sqrt{1 - (\frac{18}{64})^2}} \sqrt{(2)\left(32.2 \frac{\text{ft}}{\text{s}^2}\right) \left[\frac{(75)(144)}{62.4} - \frac{(50)(144)}{62.4} \right] \text{ft}}$$

Note that all dimensions have been converted to feet except for the A/A_1 term, which is dimensionless. When solved, the equation is

$$Q_{theor} = 7.9 \frac{\text{ft}^3}{\text{s}}$$

and

$$C_D = \frac{Q_{act}}{Q_{theor}} = \frac{5}{7.9} = 0.633$$

EXAMPLE 2

SAE 10 motor oil flows through the venturi meter in Fig. 14-8. The pressure at point 1 is 30 psig and at point 2 is 25 psig. The coefficient of discharge C of the venturi is 0.95. Calculate the flow rate.

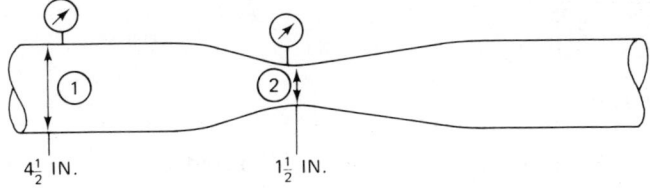

FIG. 14-8.

Solution

Equation (14-7) is used to solve the problem. The equation is

$$Q = \frac{CA_2}{\sqrt{1 - (A_2/A_1)^2}} \sqrt{2g\left(\frac{P_1}{W_s} - \frac{P_2}{W_s}\right)}$$

From Table I in the Appendix, the specific gravity of SAE 10 motor oil is 0.91. The specific weight W_s then is

$$W_s = (0.91)(62.4) = 56.7 \frac{\text{lb}}{\text{ft}^3}$$

Substituting into the equation,

$$Q = \frac{(0.95)(\pi/4)(1.5/12)^2}{\sqrt{1 - \left[\frac{(\pi/4)(1.5)^2}{(\pi/4)(4.5)^2}\right]^2}} \sqrt{(2)(32.2)\left[\frac{(30)(144)}{56.7} - \frac{(25)(144)}{56.7}\right]}$$

The A_2/A_1 term is dimensionless, so each of the two areas is left in square inches. All the other dimensions are in feet.

When the equation is simplified and solved for Q,

$$Q = \frac{0.01163}{\sqrt{1 - (2.25/20.3)^2}} \sqrt{(64.4)(76.2 - 63.5)}$$

$$= 0.34 \text{ ft}^3/\text{s}$$

QUESTIONS AND PROBLEMS

14-1.—

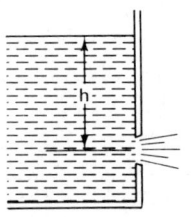

FIG. 14-9.

The orifice shown in Fig. 14-9 discharges to the atmosphere. The diameter of the orifice is 4 in. and the head h is 35 ft. If the fluid is water and the coefficients of contraction and velocity are 0.60 and 0.98, respectively, calculate the discharge.

14-2.—An orifice in a pipeline is calibrated for use with water. Assuming no change in the coefficient of discharge, what is the percent error introduced if kerosene is used instead?

14-3.—Calculate the flow in the venturi meter in Fig. 14-10. The fluid is water.

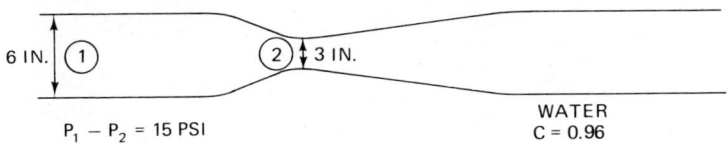

P₁ − P₂ = 15 PSI   WATER C = 0.96

FIG. 14-10.

14-4.—A pitot-static tube is used to measure the velocity of water flowing in a pipe. If the difference in the total and static pressure heads is 15 psi, what is the velocity?

14-5.—Gasoline flows through the pipeline orifice in Fig. 14-11 with conditions as shown. Calculate the flow rate.

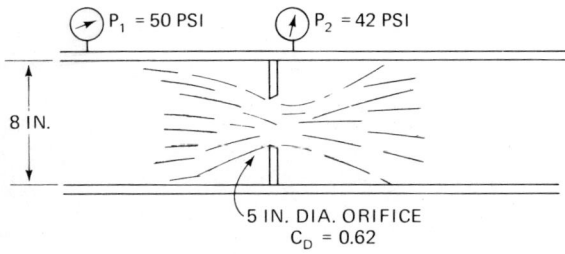

FIG. 14-11.

chapter fifteen

FRICTION LOSSES IN FITTINGS, VALVES, AND OTHER DEVICES

15-1 INTRODUCTION

In fluid power systems it frequently happens that friction losses, caused by fittings, valves, entrances, and other flow-control devices, are greater than those resulting from flow in the tubing and piping of the system. This can be visualized by examining Fig. 15-1. Here is a compact fluid power system with a minimum of length of pipe and tubing but with a total of 16 fittings, valves, and other accessories. If piping is correctly sized, it is reasonable to expect higher frictional losses from the fittings and valves than from the pipe.

Mathematical methods for estimating pressure drops caused by friction are therefore needed for fittings just as they are for piping. The methods that involve fittings and other local flow resistances are an extension of the basic empirical equations used for pipe.

15-2 MATHEMATICAL METHODS

The principle on which these methods are set up is the fact that friction losses are a function of the $v^2/2g$ term of Bernoulli's equation. This was

146 FRICTION LOSSES IN FITTINGS, VALVES, AND OTHER DEVICES / Chap. 15

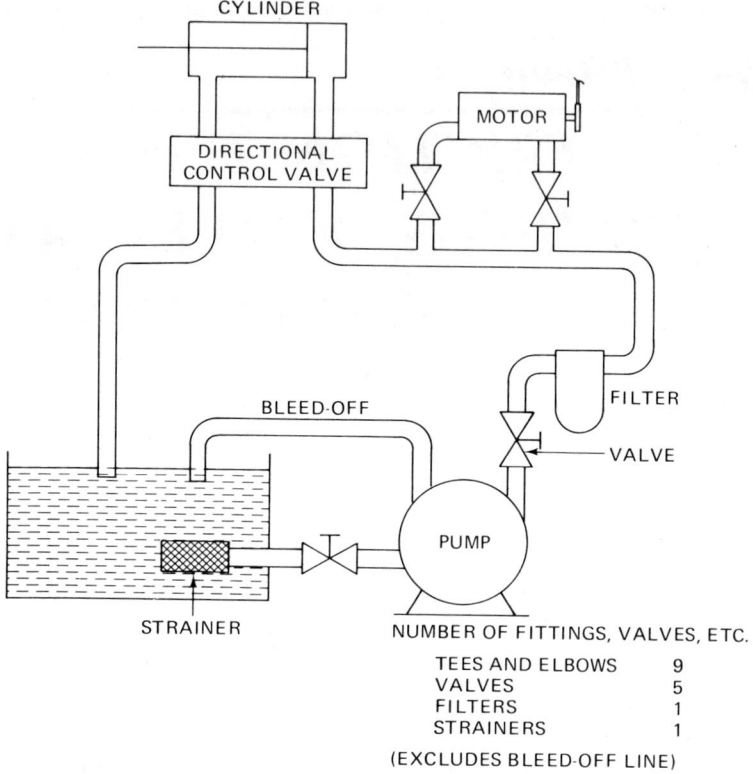

FIG. 15-1. Fluid power system with high utilization of valves and fittings.

brought out in Chapter Nine, where Darcy's equation was used to express the head loss due to friction:

$$h_L = f \frac{L}{d} \frac{v^2}{2g} \qquad (9\text{-}5)$$

This equation is used for incompressible fluid flow through pipes of various types. It is usable for both laminar and turbulent flow by varying the form the friction factor f takes.

Equation (9-5) is adapted to calculation of local flow resistances by two methods. They are:

1 / The conversion of fittings and other local flow resistances to an "equivalent length of pipe" of the same size as the fittings, valve, and other accessories. This allows calculation of the friction loss to be done in essentially the same manner as if the straight pipe were used. The friction factor f can be selected on the basis of

the Reynolds number of the fluid and calculations completed in the normal manner.

2 / The replacement of the $f(L/d)$ term by a term called the K factor, an empirical dimensionless term that describes the particular fitting or other type of flow resistance involved. The K factor comes from tests and data accumulated over periods of time by manufacturers and research organizations in the field.

15-3   EQUIVALENT LENGTH OF PIPE

Table 15-1 lists values of the L/d term from Eq. (9-5) for various fittings. The diameter and equivalent length from this term are based on Schedule 40 pipe sizes. The internal diameters for Schedule 40 pipe are listed in Table VI in the Appendix.

To determine the equivalent length of a fitting, select the correct L/d term from the table. The equivalent length then is this factor multiplied by the diameter. For example, suppose that you are to find the equivalent

Table 15-1   Values of L/d Ratios for Selected Fittings*

Fitting	L/d
Globe valve, perpendicular stem, open	340
Globe valve, Y pattern, open	160
Angle valve, open	145
Gate valve, open	13
Gate valve, $\frac{3}{4}$ open	35
Gate valve, $\frac{1}{2}$ open	160
Gate valve, $\frac{1}{4}$ open	900
Swing check valve, open	135
Check valve, open	150
Plug cock, full port, two-way, open	18
Plug cock, reduced port, three-way, straight through	44
Elbow, 90° standard	30
Elbow, 90° long radius	20
Elbow, 45° standard	16
Elbow, 90° square corner	57
Tee, standard, through run	20
Tee, standard, through branch	60
Return bend, close pattern	50

*Reproduced from Donald G. Newton, *Fluid Power for Technicians*, p. 53, with the permission of the publisher, Prentice-Hall, Inc., Englewood Cliffs, N.J.; copyright © 1971.

length of Schedule 40 pipe which corresponds to a 2-in. angle valve, fully open. From Table 15-1 the L/d term is 145, or $L/d = 145$. Then $L = 145d$. From Table VI in the Appendix, $d = 2.067$ in. Since the equivalent length is desired in feet, use $d = 2.067/12$. The equivalent length is then $(145)(2.067/12) = 25$ ft.

An example of the use of the table and the calculation of total pressure drop in the piping and fittings follows.

EXAMPLE 1

Total length of 1-in. straight pipe = 10 ft. Calculate the total equivalent length of 1-in. pipe in the system shown in Fig. 15-2.

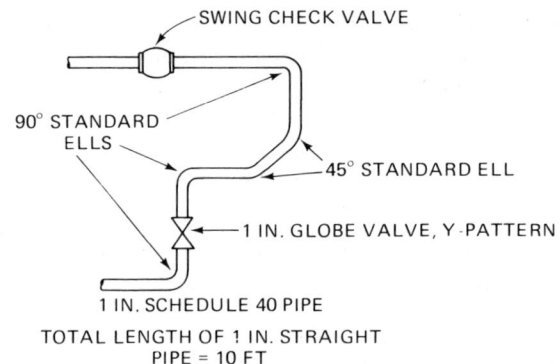

FIG. 15-2.

Solution

From Table VI in the Appendix, the inside diameter of 1-in. Schedule 40 pipe is 1.049 in. L/d values are taken from Table 15-1 for the fittings and are converted to the following equivalent lengths:

Fitting	$L/d \times \dfrac{1.049}{12} = L$		No. Fittings	Total Equivalent Length (ft)
Swing check valve	135	11.8	1	11.8
90° std. elbow	30	2.6	3	7.8
45° std. elbow	16	1.4	2	2.8
Globe valve	160	14.0	1	14.0
			Total	36.4
			+ straight pipe	10.0
			Total equivalent length	46.4

This example illustrates how important a part fittings may play in the total pressure drop of the system.

EXAMPLE 2

With the system and conditions as shown in Example 1, calculate the total pressure drop due to friction. The liquid flowing in the system is water at a Reynolds number 8×10^4. Assume a relative roughness ϵ/d value of 0.002. The velocity is 11.9 ft/s.

Solution

The procedure now becomes the same as that used in Unit 9 for Moody's chart. From Fig. 9-4, using Reynolds number and relative roughness values above, enter the chart and find the friction factor f of 0.026. Equation (9-5) can then be used as follows:

$$h_L = f \frac{L}{d} \frac{v^2}{2g}$$

The length L here is the total equivalent length of 46.4 ft from Example 1. The diameter d is the diameter of the pipe, 1.049 in. The velocity is given, so the values may be substituted in the equation.

$$h_L = (0.026) \frac{46.4}{1.049/12} \frac{(11.9)^2}{(2)(32.2)}$$

$$= 30.3 \text{ ft of water} = (30.3)(0.433) \text{ psi}$$

$$= 13.3 \text{ psi}$$

If the Reynolds number and relative roughness values had not been given, they would have had to be calculated using the procedures in Chapter 9.

15-4   K FACTOR

Use of the K-factor method requires only that the factor be known for the valve, fitting, or other local resistance used in the system. The data supplied by manufacturers and researchers for the K factor are, in most cases, accurate. It is empirical data, however, and most error comes from its misapplication. Thus, care should be used to determine that the conditions where it is applied are as close as possible to the stated conditions that the value of K covers.

The K factor multiplied by the $v^2/2g$ term gives the head loss in feet of the fluid. It can be converted to pounds per square inch by the usual methods of using the specific weight of the fluid. Table 15-2 shows the values of the K factor for various local resistances.

Table 15-2 Conduit Branches in Tubing

ENTRANCE LOSSES

- K = 0.5 FROM DULL TO SHARP EDGE
- K = 0.68 TO 2.5 FROM BROKEN TO RAZOR SHARP EDGE
- 90°, K = 0.25
- 60°, with table:

X/d	K
0.10	0.18
0.15	0.15
0.25	0.13
0.40	0.10

- K = 0.1 WHEN R/d = 0.12
  K = 0.06 WHEN R/d = 0.16
- K = 0.04 WHEN WELL ROUNDED

CONDUIT BRANCHES IN TUBING

ARROWS IN OPPOSITE DIRECTION INDICATE K HOLDS FOR BOTH FLOW DIRECTIONS

- K = 1.2
- K = 0.1
- 45°, K = 0.5
- 45°, K = 2.5 TO 3
- 45°, K = 0.06
- 45°, K = 0.15
- 90°, K = 1.12

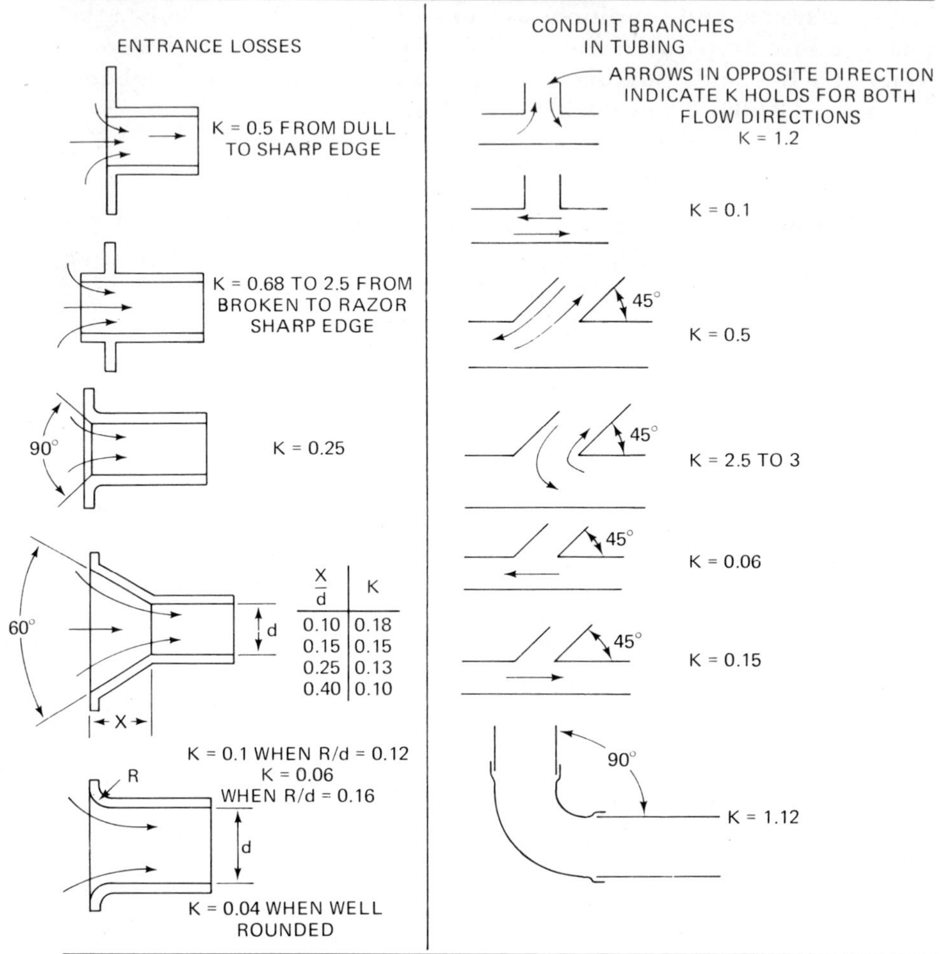

Reproduced from Franklin D. Yeaple, ed., *Hydraulic and Pneumatic Power and Control*, with the permission of the McGraw-Hill Book Company, New York.

EXAMPLE

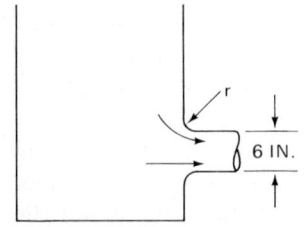

FIG. 15-3.

The entrance from the reservoir in Fig. 15-3 into a 6-in. line is to be designed for minimum losses. What should the radius r be?

Solution

Referring to Table 15-2, a value of K of 0.06 is listed when $r/d = 0.16$. This is a small value and the losses will be equal to $0.06\,(v^2/2g)$ if the radius is designed according to this ratio. The calculated value of the radius is $(0.16)(6) = 0.96$. Therefore, a design value of 1 in. would be satisfactory for the radius.

QUESTIONS AND PROBLEMS

15-1.—The header in Fig. 15-4 is made of tubing and is connected on the output side to three smaller tubing lines at A, B, and C. The total flow rate into the header is 0.6 ft^3/s. The 1-in.-diameter sections of tubing have equal quantities flowing through them. The $\frac{1}{2}$-in.-diameter tubing has one fourth the flow quantity of a 1-in.-diameter tubing line. With oil at a specific gravity of 0.84, calculate the entrance losses at A, B, and C.

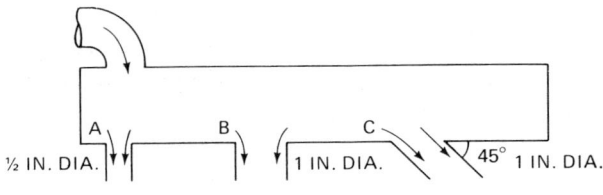

FIG. 15-4.

15-2.—Calculate the equivalent length of Schedule 40 pipe for the following:

a—A 2-in. 90° standard elbow

b—A 1-in. gate valve, half-open

c—A 3-in. open, perpendicular stem globe valve

15-3.—Calculate the friction loss in a 2-in. perpendicular stem globe valve, fully open, with water flowing through it at a velocity of 40 ft/s. Use the roughness factor for steel pipe when determining the friction factor.

15-4.—Calculate the friction loss from all the fittings in the system in Fig. 15-5. All elbows are 90° standard and the piping is $1\frac{1}{2}$-in. Schedule 40 steel pipe. The pump circulates 20 gpm of kerosene at a temperature of 90°F through the system.

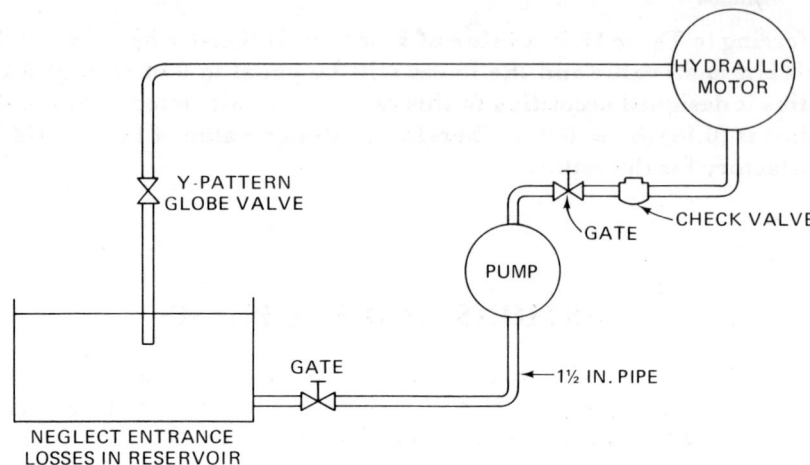

FIG. 15-5.

15-5.—Design a piping system using 2-in. Schedule 40 pipe to distribute solvent (specific gravity of 0.8) from the storage tank in Fig. 15-6 to three headers. Distribution is by gravity flow and it is important to keep friction losses to a minimum. Sketch the system, showing all fittings and valves. Calculate the pressure drop caused by the fittings and valves. Provide cutoff valves for each header.

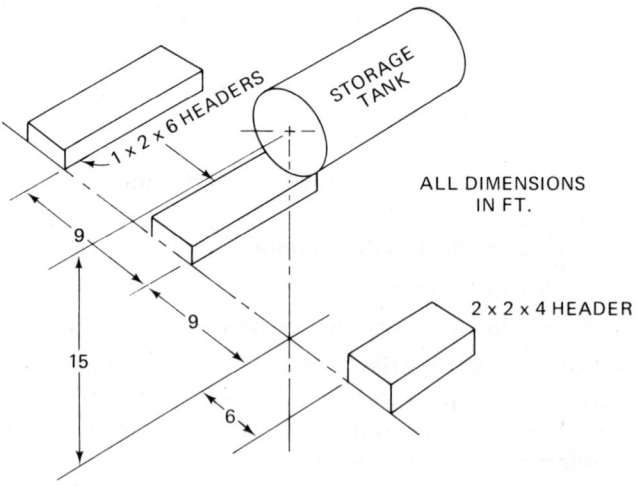

FIG. 15-6.

chapter sixteen

FLUID POWER

16-1 INTRODUCTION

The principles of fluid mechanics as outlined in the preceding chapters are used in many different fields. The civil engineer uses them to design water supply and sewage systems for cities. The aeronautical engineer adapts fluid mechanics to the study of lift and drag for aircraft. For example, a popular type of wing section for aircraft is called a *laminar flow wing* because it uses the principle of laminar flow to reduce drag.

One of the largest and most important fields which depends on the application of fluid mechanics is fluid power. *Fluid power* might be defined as the application of the principles of fluid mechanics to the design of products that perform mechanical work. This definition is a little too broad, since such applications as the water wheel, to produce electrical power, are not in the fluid power field. However, for products such as actuators, motors, and cylinders, the definition is adequate.

This definition of fluid power can be expanded upon by examining the applications shown in Figs. 16-1 through 16-8. These illustrations show the use of fluid power in different industries and different commercial fields. The most significant thing brought out by the illustrations is the fact that fluid power is important in such diverse fields.

AGRICULTURE
HYDRAULIC POWER PACKAGE
IMPLEMENT CYLINDER

Fluid power lifts the plow from its furrow, and lowers it for the next, as the farmer moves on his mechanized way. It controls the business ends of harrows, harvesters, hullers, planters, threshers, seeders, pickers, and diggers. And, long before any such equipment arrives at the farm, its manufacturer will have made liberal use of fluid power in the shops and machines which fashioned it.

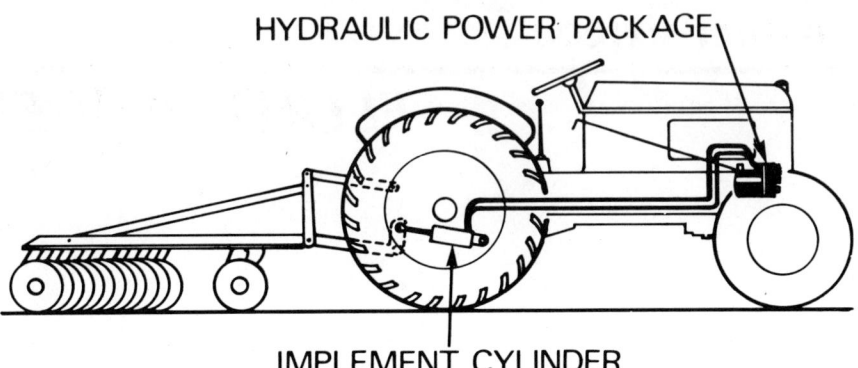

AUTOMOTIVE
POWER BRAKES
POWER STEERING

Power brakes and power steering are typical fluid power devices. Automatic hydraulic transmissions rate as a high achievement of the art. Back at the production lines, fluid power operates the presses which form body parts and fenders, punches holes and heads rivets to hold the frame together, actuates heads, slides and chucks of the machinery producing engine parts, spins the wheels which polish the finishes and may even index "the line" itself.

FIG. 16-1. (Figs. 16-1–16-8 are *reproduced with permission of* the Fluid Power Foundation.)

Sect. 16-1 / INTRODUCTION

AVIATION

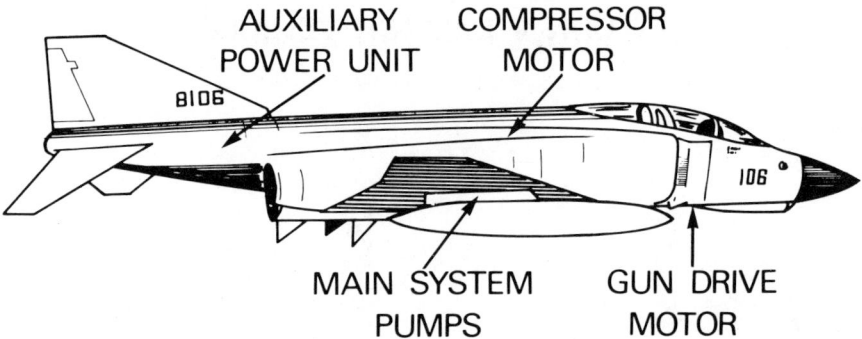

Your pilot tucks up his landing gear and controls ailerons, rudders, elevators and trim tabs with fluid power. In the fighting ship, it's fluid power which opens the bays and rotates the turrets. "One-shot" casting operations for light aluminum and magnesium parts need fluid power to close the dies and inject the metal. Wing and fuselage sections are formed in fluid power operated stretch presses.

CONSTRUCTION

When it comes to moving dirt or rock—for roads, tunnels, dams or seaways—look for fluid power at the member which bites the earth. You'll find it on road graders, shovels, crushers, and rock drills. Rollers, scrapers, bulldozers, vibrator screens, draglines, truck loaders and asphalt mixers all use it. Even the smaller tools of construction get into the act—heading rivets, breaking concrete or jacking up structural members.

FIG. 16-2.

AEROSPACE
SATURN V LAUNCH VEHICLE

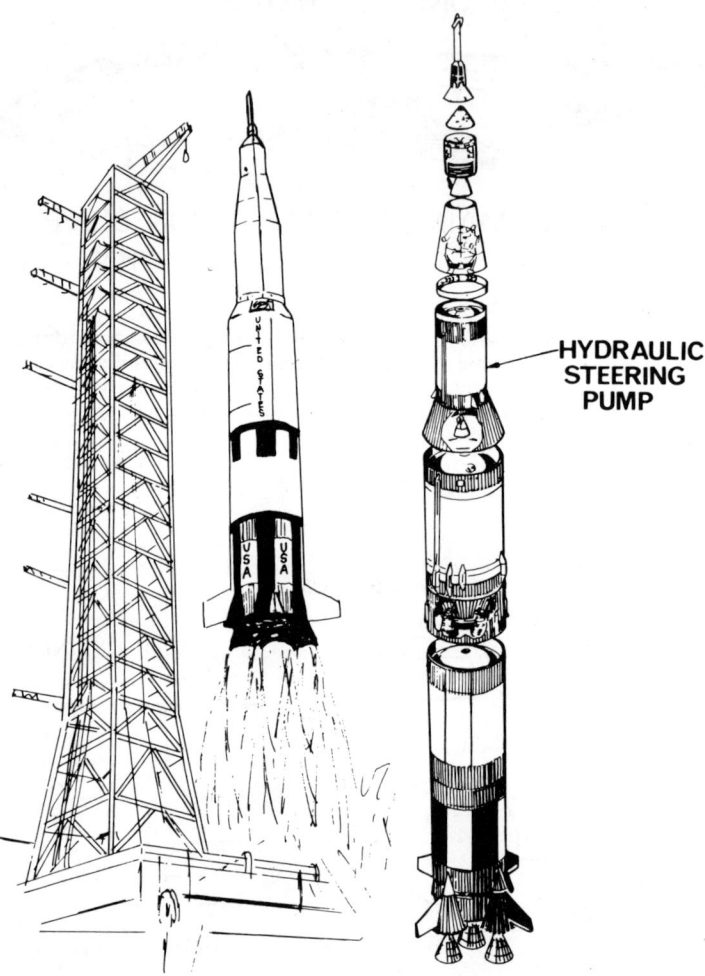

Gigantic rockets that hurl satellites into orbit around the Earth, and that carry men to the moon and other planets, depend on fluid power to control their flight. Only fluid power systems have the "guts" and power to control, with the delicacy of a feather touch, the millions of horsepower released by rocket engines and direct the payload to its destination.

FIG. 16-3.

Sect. 16-1 / INTRODUCTION

MATERIAL HANDLING

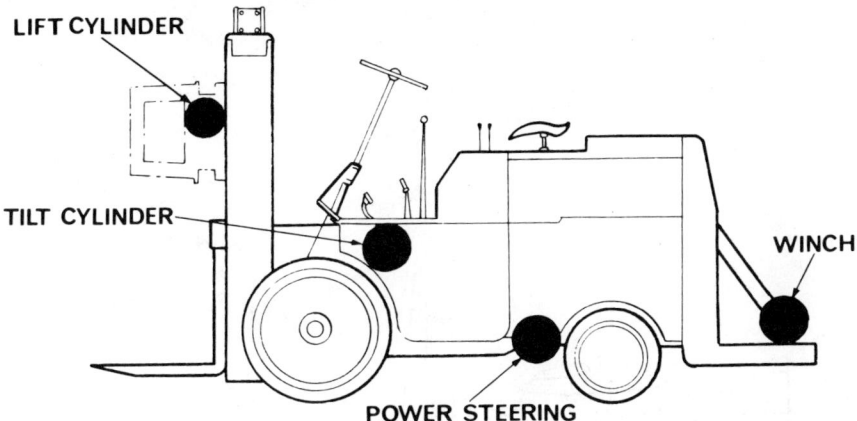

The fork lift trucks which move so efficiently and in such large fleets throughout almost all plants and warehouses these days are so versatile they can almost talk. Their telescoping masts are fluid powered. So are their grab jaws and pusher bars. Conveyors, hoists, cranes, dumpers, tilting ramps and leveling docks are but a few others in the roster of fluid powered tools of modern materials handling.

PLASTICS

PLASTIC EXTRUDER

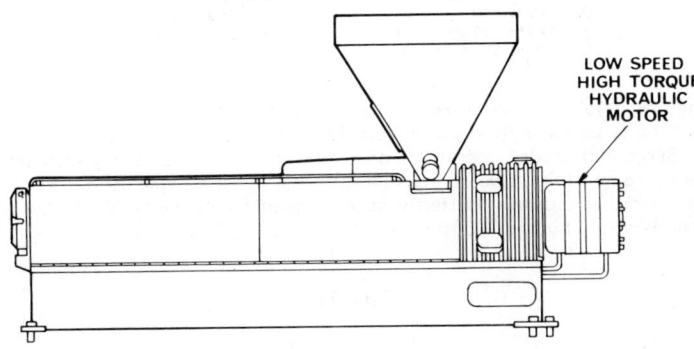

Plastics fabrication is a molding business. Injection, preform, laminated and vacuum molding pressures are all exerted and controlled by fluid power systems. On an injection machine, for example, the dies are closed and the plastic material forced into their cavities, by fluid power. Extrusion presses for tubular plastic products (like garden hose) also use fluid power for the push.

FIG. 16-4.

MARINE

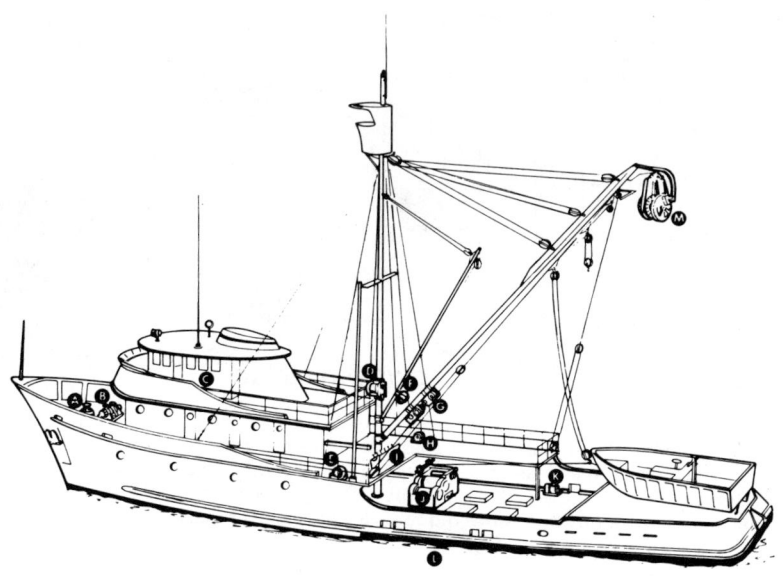

- Ⓐ CORKLINE WINCH
- Ⓑ ANCHOR WINCH
- Ⓒ POWER STEERING
- Ⓓ TOPPING WINCH
- Ⓔ VANG WINCH
- Ⓕ BRAILING BOOM WINCH
- Ⓖ MAIN BOOM WINCHES
- Ⓗ VANG WINCH
- Ⓘ WINCH CONTROL CONSOLE
- Ⓙ PURSE SEINE WINCH
- Ⓚ CHOKER WINCH
- Ⓛ STABILIZER FIN
- Ⓜ PURETIC POWER BLOCK

One of the classic jobs of fluid power in marine shipping is automatic helmsmanship. Cargo handling is partly a fluid power job. Even the hatch covers answer to such acluation. Barge and dredge operations on inland waterways are being increasingly mechanized through fluid power. Canal and river lock routine make good use of fluid power for operating capstans, butterfly valves, winches and hoists. Much dock and shipyard machinery is hydraulic or pneumatic.

FIG. 16-5.

Sect. 16-1 / INTRODUCTION

FIG. 16-6.

Before paper becomes paper it will have passed through many stages of continuous web flow from the pulp form. The amount of roll pressure exerted on the paper stock as it is consolidated from mat to finished stock is always critical. Controls of these pressures, together with speed control are perhaps fluid power's most important applications in paper-making. Calender roll adjustment is typical. Throughout paper mills you'll find other fluid power devices—in drives, feeders, reels, drivers, shredders, vibrators, sizers, coating and laminating units, hydropulpers and finishing tables.

PRINTING

ROTOGRAVURE PACKAGE PRINTING PRESS

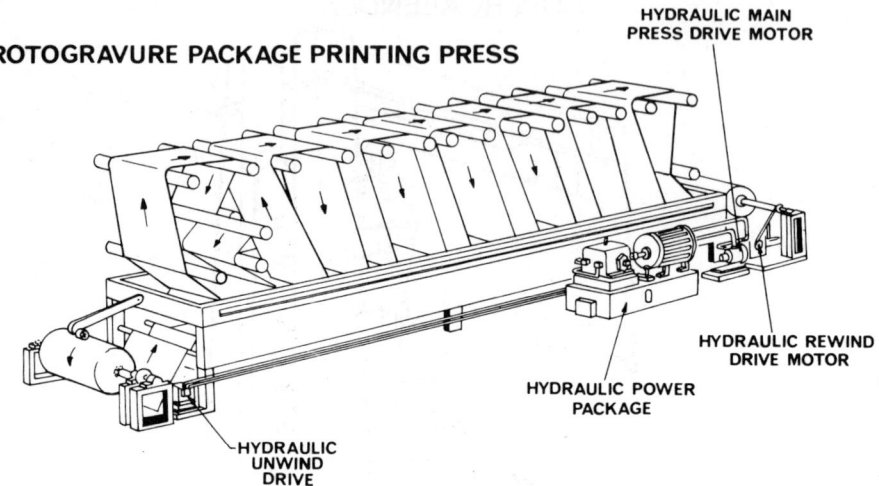

Today's printing presses operate at new lightening speeds. Yet these speeds must be closely adjustable over wide ranges and respond instantly to push-button cues. Fluid power variable speed drives and controls made it possible. Inching, acceleration, decelerating, breaking, roll pressure, and unwind and rewind tensioning are accurately performed the fluid power way. Fluid power eliminates jerking, one cause of those occasional off-register pictures you used to see in earlier color pictures.

FIG. 16-7.

RUBBER

RUBBER PREFORMING MACHINE

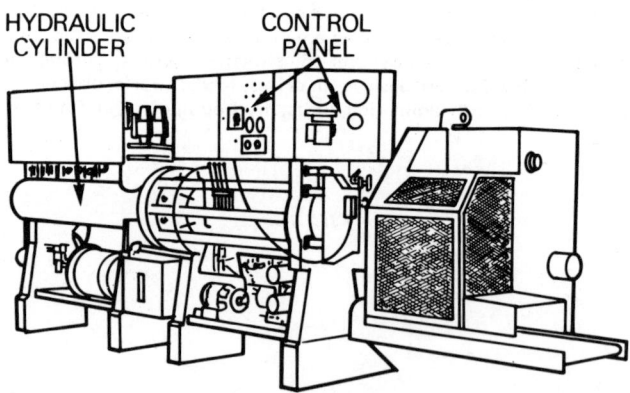

Fluid power shows up in lots of places in the rubber business—in the production of the new synthetics, in compounding, and in the fabrication of finished products. Roll pressure control during compounding is a major need (squeezing each batch time and again between the rolls of a mixer).

FIG. 16-8.

16-2 HYDRAULICS AND PNEUMATICS

Fluid power applications are hydraulic, pneumatic, or combinations of the two. Hydraulic applications are those utilizing incompressible fluids (liquids). Pneumatic applications use air for the fluid. Compressible fluids other than air may also be used, but the term "pneumatics" is restricted to air. Combinations of hydraulic–pneumatic systems are used also. These are called *air–oil systems*. The hydraulic lift at the service station is an air–oil system.

16-3 SPECIAL APPLICATIONS

The applications of the principles of fluid mechanics to industry are limited only by the ingenuity of the designer. The applications shown in Figs. 16-1 through 16-8 are well-engineered products for commercial use and space research. There are also many special applications not as well known which have been developed over the years to perform a specific function in manufacturing. Three of these are shown in Figs. 16-9, 16-10, and 16-11.

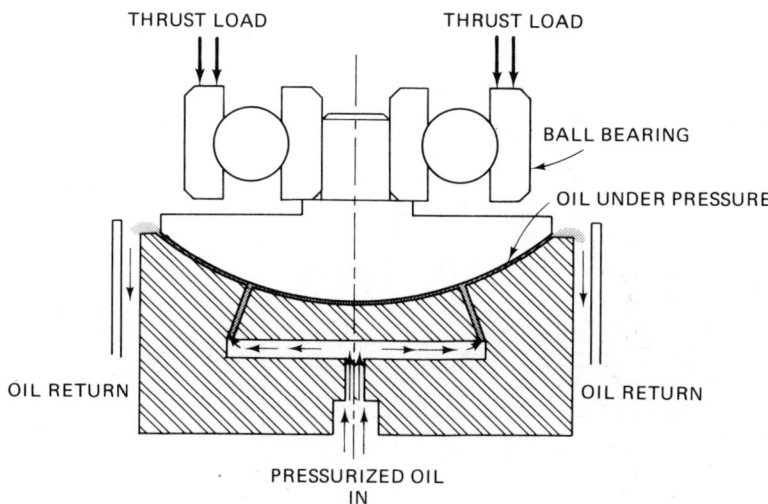

FIG. 16-9. Hydrostatic self-aligning spherical bearing.

In Fig. 16-9 is shown a self-aligning spherical thrust bearing supported by an oil film. Pressure is maintained on the oil by a pump (not shown). The oil, which continually leaks out of the bearing, drops down into a cavity and is returned to the reservoir. The design makes use of the viscosity of the fluid and its load-carrying ability, represented by the equa-

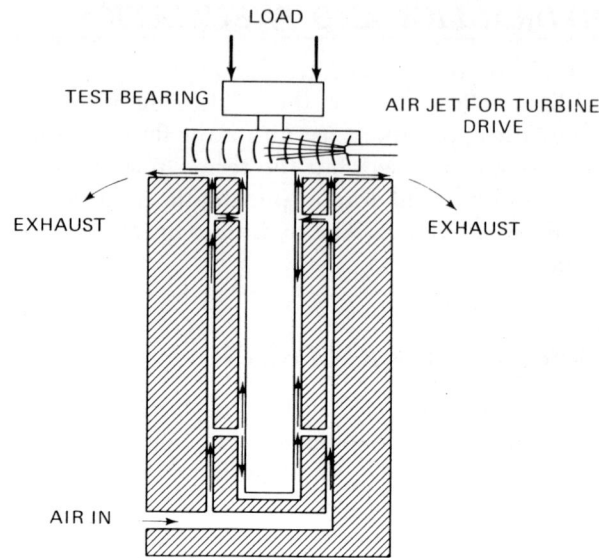

FIG. 16-10. Spindle supported by hydrostatic air bearing.

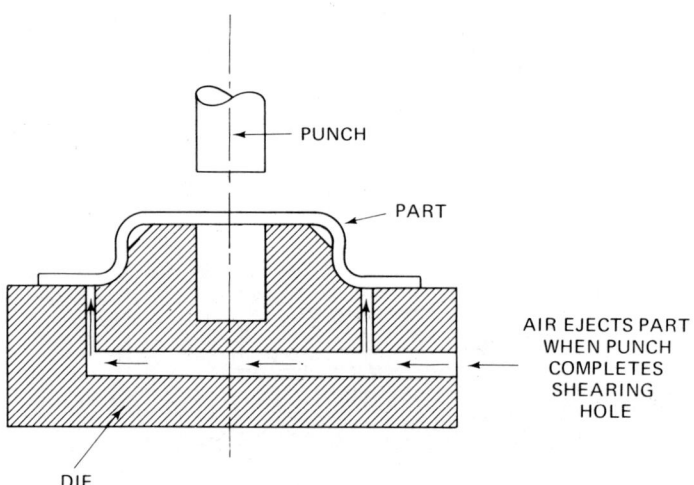

FIG. 16-11. Die and punch with automatic ejection of part by air.

tion $F = PA$. The bearing is part of a gage to measure ball bearings with a high degree of accuracy under a thrust load. It was developed by a ball-bearing manufacturer.

The precision spindle in Fig. 16-10 is supported by a hydrostatic air bearing. The low viscosity of air provides very little friction to resist rotation. The air is pressurized to provide the load-carrying capacity for radial and thrust loads. The principle of impulse–momentum is used to rotate the spindle, using a jet of air impinging on the turbine.

Air ejection of parts after completion of a manufacturing operation is a typical application lending itself to individual design efforts. An example of how this works is shown in Fig. 16-11. The air circuit is designed into the die and at the completion of the operation the part is automatically ejected.

QUESTIONS AND PROBLEMS

16-1.—Give an example of an air–oil fluid power system.

16-2.—Give two fluid power applications in the automobile.

16-3.—What is a pneumatic fluid power system?

16-4.—What is a hydraulic fluid power system?

chapter seventeen

SELECTING THE FLUID POWER SYSTEM

The choice of what type of fluid power system to use must be made in designing a machine. In general, this means it must be decided whether to use hydraulics, pneumatics, the combination air–oil system, or both separate hydraulic and pneumatic systems. This chapter discusses the advantages of the different types of systems and where each may be used to the best advantage.

17-1 THE HYDRAULIC SYSTEM

The most frequently used actuator in fluid power systems is the cylinder. The operation of a double-acting cylinder is shown in Fig. 17-1. Although cylinder types are explained in Chapter Eighteen, the cylinder shown here can be used to explain some of the differences between hydraulic and pneumatic systems.

In the hydraulic cylinder the fluid is incompressible. If a given quantity of fluid enters the cylinder, the volume of the cylinder at this end must increase in the same amount. The only way for this to occur is for the piston to move out a distance sufficient to provide the increased volume.

Consider the same cylinder with air, a compressible fluid. Air enters

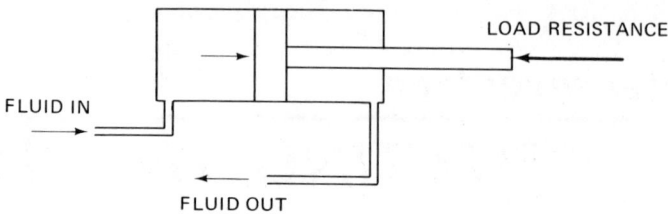

FIG. 17-1. Double-acting cylinder.

the cylinder as before. The load resistance now becomes high enough so that the piston stalls against it. The air in the cylinder has been compressed, raising the pressure to a higher value. This difference between hydraulic and pneumatic power is one of the major factors to evaluate in making the choice between hydraulic or pneumatic operation.

Some of the other characteristics of hydraulic versus pneumatic systems are as follows:

1 / **The hydraulic system normally requires its own self-contained system, including pumps and reservoirs. Figure 17-2 illustrates a typical circuit.**

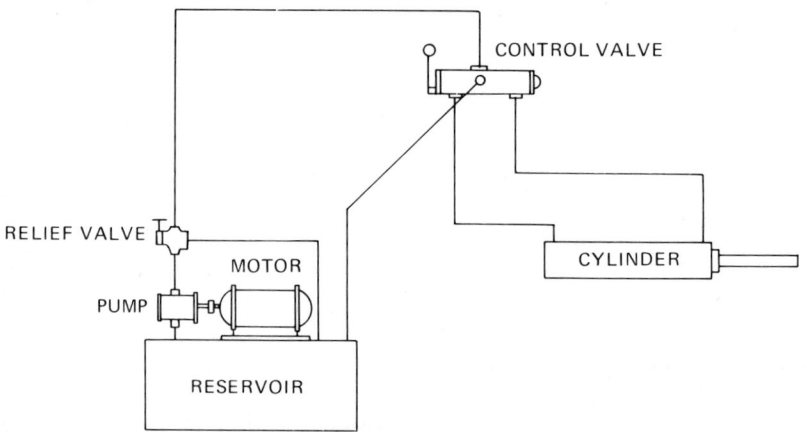

FIG. 17-2. Hydraulic circuit.

2 / **Higher-force outputs can be obtained from hydraulic systems than from pneumatic ones.**

3 / **The exhaust of components in the hydraulic system has to be returned to the reservoir; in pneumatic systems the exhaust is to the atmosphere.**

4 / **The velocity of the cylinder piston rod can be determined from** the flow rate of the hydraulic fluid. Because of the positive displacement characteristics resulting from the incompressible fluid, the velocity is constant if the flow rate is constant. The velocity of the pneumatic piston varies if the load resistance varies.

5 / Accurate positioning can be obtained by hydraulics but not with air.

17-2   PNEUMATIC SYSTEM

Some advantages of the pneumatic system are as follows:

1 / A pneumatic system is cleaner than a hydraulic one because of the lack of oil leakage.

2 / Where loads are sufficiently light, cylinder operation and response times are frequently faster than that obtained with hydraulics.

3 / Most of the time the compressed air supply is already available from a shop air source, eliminating the need for its own self-contained energy source.

The design of a hydraulically operated tensile tester is shown in the drawing in Fig. 17-3. The tensile tester is used in metallurgical laboratories and testing laboratories to perform tensile, compression, shear, and bending tests for steels, other metals, and nonmetals. In operation the table is moved upward by the piston, in turn moving the upper head up. The lower head is stationary during the test period but can be moved up and down by the crosshead motor for adjustment prior to testing the specimen. In tension testing, the specimen is tested to failure when the upper head pulls away from the lower head. Compression, shear, and bending test specimens are mounted between the table and lower head. The machine capacity is 120,000 pounds, or 60 tons. The selection of a hydraulic system, then, is dictated by the high force output of 60 tons required by the piston.

Note the application of the Bourdon tubes at the left. These are used to sense the pressure in each of the three force ranges used (only one range is used at a time). The movement of the tip of the Bourdon tube is sensed electrically by a differential transformer, amplified, and displayed on the front panel as a force.

A design where air is necessary is that of the air springs shown in Fig.

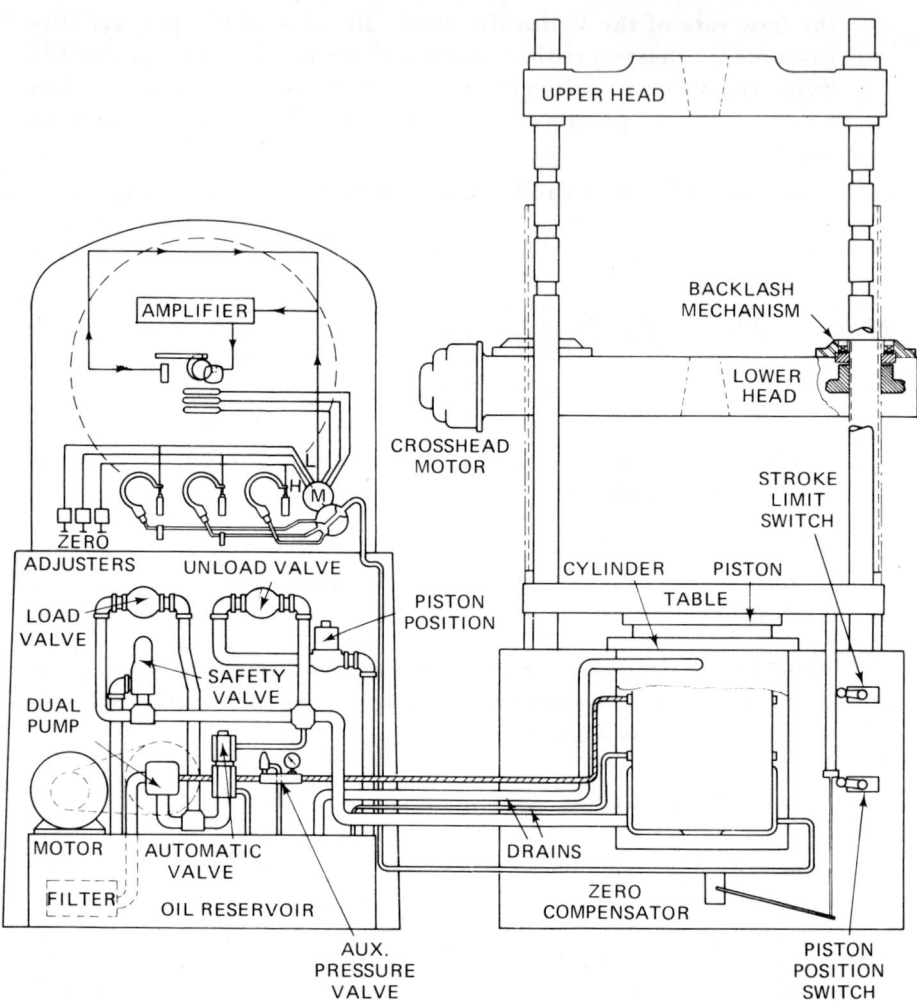

FIG. 17-3. Tinius Olsen super L hydraulic universal testing machine. (*Reproduced by permission of* Tinius Olsen Testing Machine Co., Inc.)

17-4. This is the air suspension system for a vehicle. Air suspension is used on buses and other heavy vehicles. Because of its compressibility, air functions as a spring in the suspension system. Thus, the success of the design is due to the compressibility of the fluid used. Unlike most pneumatic systems, there is no external source of compressed air available, so a compressor has to be provided as part of the system.

Sect. 17-3 / AIR–OIL SYSTEM

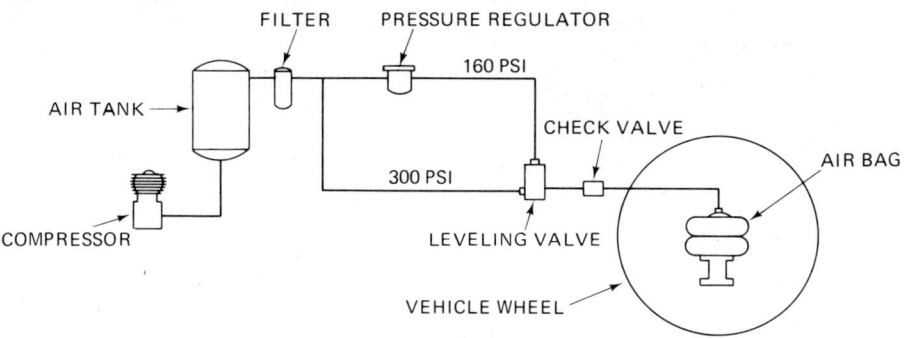

FIG. 17-4. Automotive air suspension system.

17-3   AIR–OIL SYSTEM

Combination air–oil systems provide some special advantages for some applications. The air-power lift shown in Fig. 17-5 is in reality a hydraulic lift operated by air. The air–oil system of Fig. 17-6 uses hydraulic feed on one side of the cylinder for close control of the movement and air on the other side of the cylinder for fast retraction with no load.

The air-powered lift design is used for automobile lifts. Its main advantage is that it allows operation of the hydraulic lift from the shop air supply, eliminating a separate pump and reservoir. The hydraulic fluid is forced back in the air–oil tank by the weight of the ram when air pressure is taken off.

The system in Fig. 17-6 provides for metered flow of the hydraulic fluid, which gives control of the speed with which the piston rod moves out

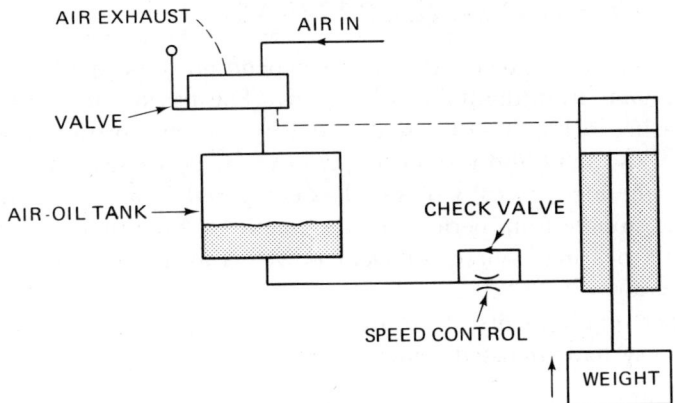

FIG. 17-5. Air-oil lift.

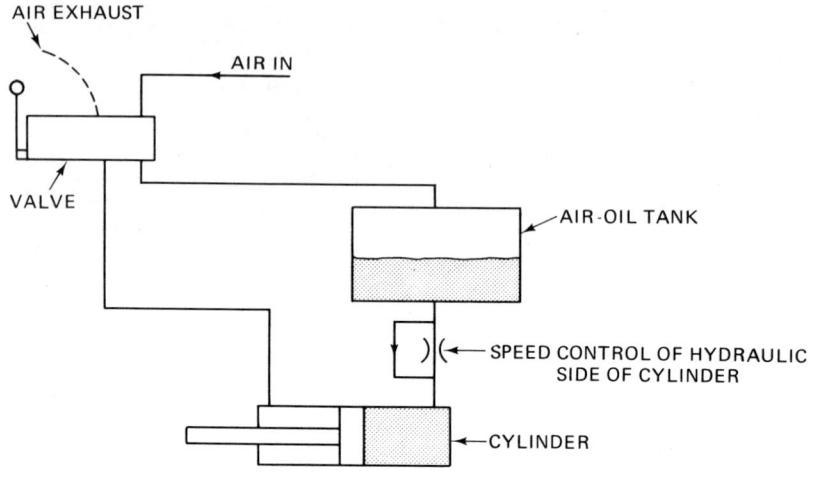

FIG. 17-6. Air-oil powered cylinder.

from the cylinder. On the return stroke the piston is operated directly by the air. There is no control over the speed of the piston on the return stroke.

Air-oil systems require careful design to prevent excessive amounts of air dissolving in the oil. The air-oil tank should be mounted at a higher elevation than the rest of the system, to allow excess dissolved air to escape from the oil.

17-4 FLUID CONSIDERATIONS

The choice of a fluid power system may depend on some special characteristic obtainable from the fluid to be used. The use of air for the hydrostatic bearing in Fig. 16-10 was dictated by the low torque available for rotation of the shaft and the sensitivity of the measurements taken on the spindle. The high viscosity of an oil as compared to air would have prevented the spindle from performing its function. Note that oil is used for hydrostatic bearings where sufficient power is available to overcome the higher friction.

Other requirements of an application that may help determine the choice of a system are listed below:

1 / Lubrication: the lubrication available from most hydraulic fluids may be desired in the functioning of the machine, whereas if air were used, excessive wear would result.

2 / Corrosion resistance: mineral-oil-base hydraulic fluids provide some corrosion resistance. Air has to have all traces of moisture removed to prevent it from causing corrosion; thus, hydraulic fluids are safer in this respect.

3 / Fire resistance: most hydraulic fluids are flammable; therefore, air is usually a better choice for fire resistance.

17-5 CONDITIONING OF THE FLUID

The fluid used in the hydraulic or pneumatic system must be conditioned properly to obtain satisfactory operation of the system. In most cases the conditioning consists of filtration of the fluid and removal of moisture. Cooling may be required for the hydraulic fluid. It is common practice to provide an oiler to provide lubrication to an air supply.

Figure 17-7 illustrates a hydraulic reservoir with pump, piping, a strainer, and a filter. The primary fluid conditioning required for a hydraulic system is filtration. The strainer provides filtration of coarse particles on the suction side of the pump, and additional filtration is provided by the filter located on the output side of the pump. The selection of the filter is dictated by the allowable pressure drop of the filter and the fineness to which filtration is desired.

A secondary conditioning that may be required of hydraulic fluid is cooling. Excess heat may build up in the system and it is sometimes necessary to provide a heat exchanger to remove it. The reservoir itself provides some function as a heat exchanger, particularly if its capacity is larger with respect to the system capacity.

The conditioning required for air is different than that for hydraulics.

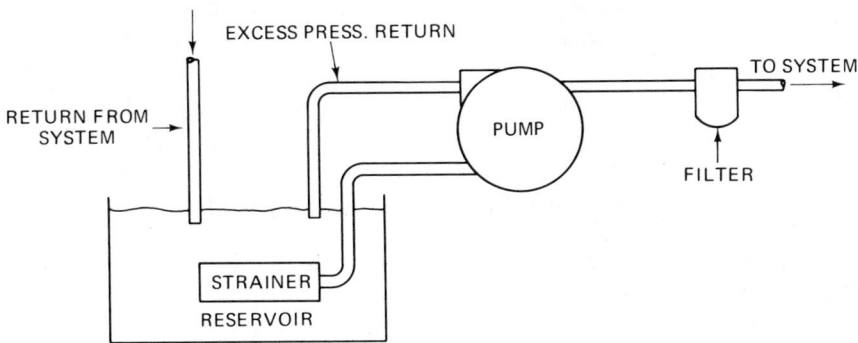

FIG. 17-7. Strainer and filter in hydraulic circuit.

Figure 17-8 illustrates a typical unit consisting of filter, pressure regulator, and lubricator. These are frequently used together as shown, to condition the air from a shop system prior to being used in the fluid power unit. Filtration and removal of all water and other liquids is particularly important. Most shop air systems build up sufficient water in the piping from condensation to cause considerable damage from corrosion if not removed. Added to this is usually sludge and oil from the compressor.

The lubricator adds clean, metered amounts of oil using a wick to feed the oil. Sometimes it is necessary to reduce the relative humidity of the air. Since the filter will remove only solids and liquids, it is necessary to add a dryer to remove the water vapor.

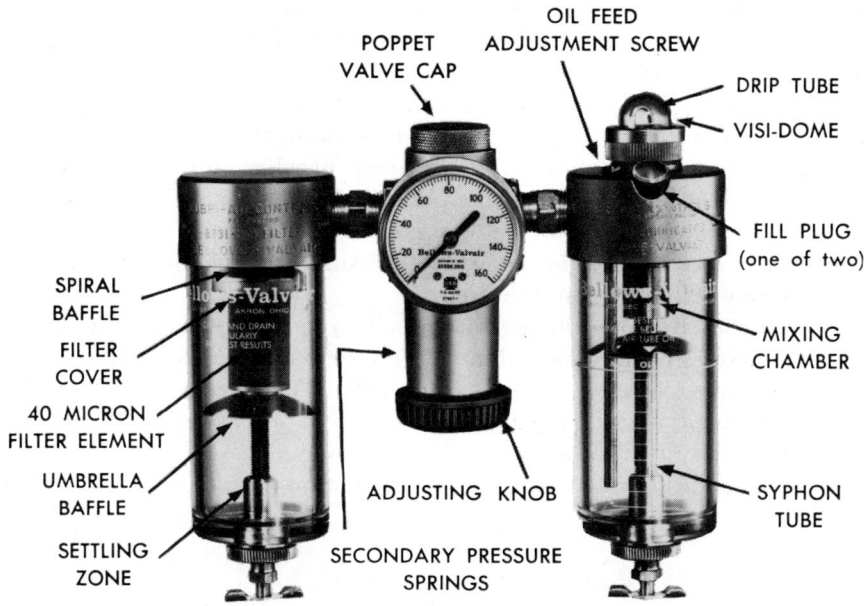

FIG. 17-8. Filter, regulator and oiler for air supply. (*Courtesy* Bellows-Valvair, Division of IBEC)

QUESTIONS AND PROBLEMS

Select the type of fluid power system that appears to be best suited for each of the applications following. Write a short paragraph justifying your choice and supporting it with all the data that you can obtain.

> *17-1.*—The 3-ton lift in Fig. 17-9 is to be built into a floor of a warehouse. Shop air at 100 psi is available, as is electrical power at 110 V.

QUESTIONS AND PROBLEMS

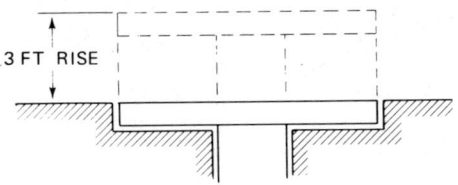

FIG. 17-9.

17-2.—The shop vise in Fig. 17-10 is to be operated by fluid power. Shop air is available.

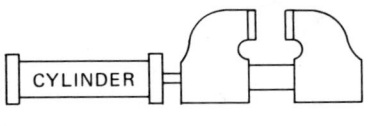

FIG. 17-10.

17-3.—The hoist linkage in Fig. 17-11 is actuated by a cylinder. The hoist capacity is 5 tons. The hoist is located in a fabricating shop and shop air is available.

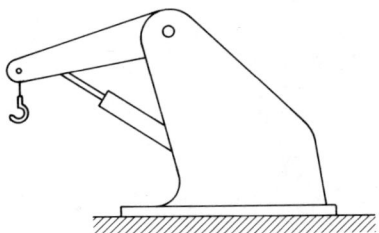

FIG. 17-11.

chapter eighteen

FLUID POWER COMPONENTS—CYLINDERS

18-1 INTRODUCTION

The work of the fluid power system is done by the actuator. Two actuator types that are of importance are the linear actuator and the rotary actuary. The cylinder is the most prominent example of the linear actuator. Rotary actuators, of which the motor is one example, are covered in another chapter.

The cylinder is seen in every conceivable fluid power application. Cylinders are produced in a variety of types and sizes for the different applications they are used in. The cylinders in Fig. 18-1 show only a part of the size range covered.

18-2 HOW THE CYLINDER FUNCTIONS

The theory of the cylinder has been touched upon in other units and is mentioned briefly here. In the single-acting cylinder of Fig. 18-2 fluid under pressure is admitted to the chamber and causes the piston and rod to move outward. If there is a resisting load on the end of the rod, the force exerted by the piston on external load is the pressure multiplied by

FLUID POWER COMPONENTS—CYLINDERS / Chap. 18

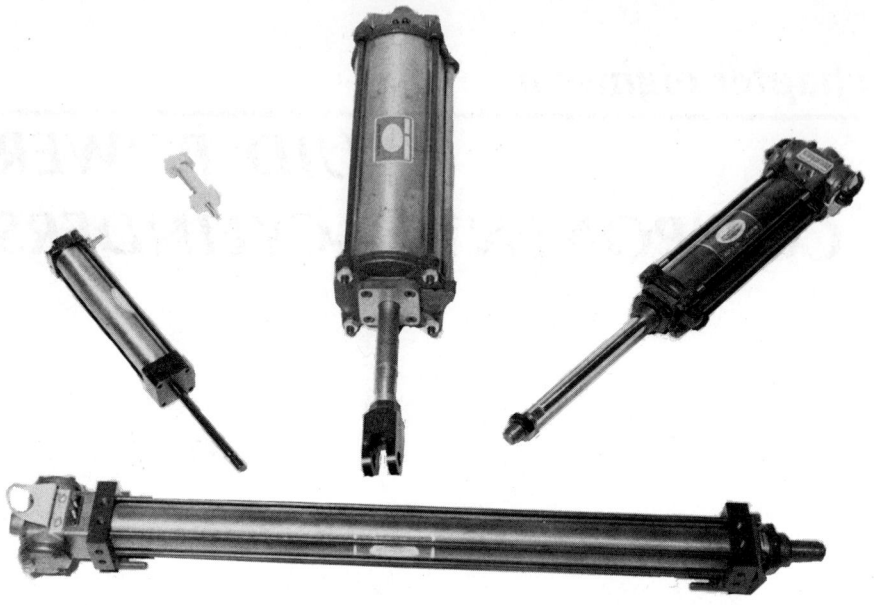

FIG. 18-1. Some common cylinder sizes.

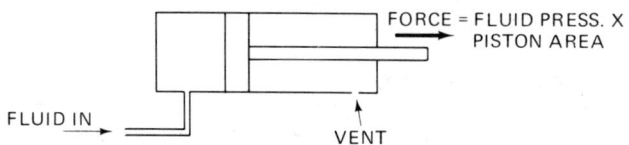

FIG. 18-2. Single acting cylinder.

the area of the piston. This is simply an application of Pascal's law, which was discussed in Chapter Two.

The force determined by this law is independent of fluid type and compressibility. The accurate positioning of a piston in midstroke can only be done with an incompressible fluid, however. Accurate control of velocity of the piston also requires an incompressible fluid. Both of these characteristics were discussed in Chapter Seventeen. Positioning of a piston can be visualized by examining Fig. 18-3.

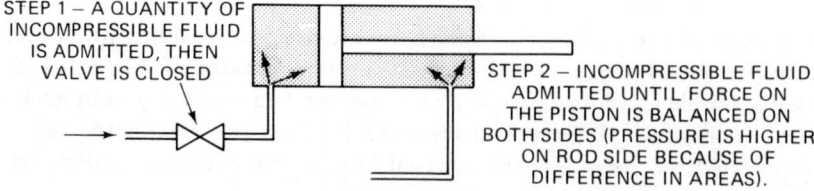

FIG. 18-3. Positioning of a hydraulic piston.

18-3 CYLINDER TYPES

Primary cylinder types are shown in the following paragraphs, together with a description and discussion of each type.

Single-Acting, Spring Return, Push Type

This cylinder type (Fig. 18-4) delivers power only on the push stroke. Mechanical spring returns rod to position.

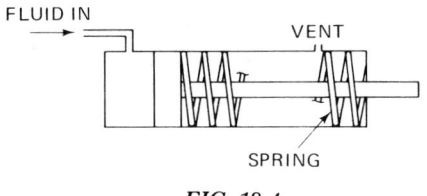

FIG. 18-4.

Single-Acting, Spring Return, Pull Type

This cylinder type (Fig. 18-5) delivers power only on the pull stroke. Mechanical spring returns rod to original position.

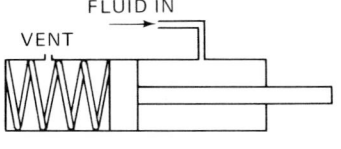

FIG. 18-5.

Double-Acting Type, Single Rod End

This type of cylinder (Fig. 18-6) is one of the most common. The piston is moved alternately in each direction by the fluid.

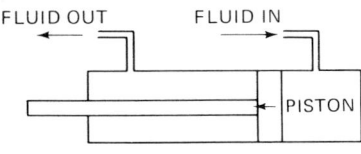

FIG. 18-6.

Double-Acting Type, Double Rod End

This cylinder (Fig. 18-7) has the same action as the single rod, double-acting cylinder but has the rod extending from both ends.

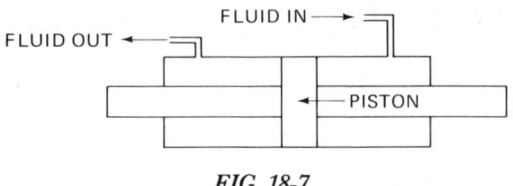

FIG. 18-7.

Single-Acting Ram Type

Thrust for the ram (Fig. 18-8) is calculated from the net area of the rod where it passes through the cylinder. The ram is used for jacks, high-tonnage actuators, presses, and other applications of this type. Return may be gravity or another external source of energy.

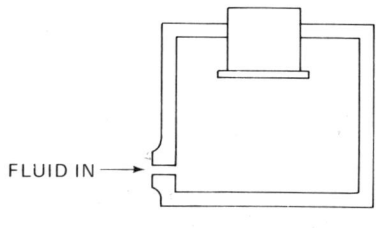

FIG. 18-8.

Telescoping Cylinder Type

This cylinder (Fig. 18-9) is used where longer lengths are required than can be obtained with standard cylinders.

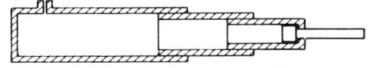

FIG. 18-9.

Diaphragm Type

The diaphragm type (Fig. 18-10) has a fairly long stroke. Return is by spring.

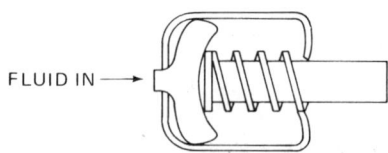

FIG. 18-10.

Sect. 18-4 / CYLINDER MOUNTING

Typical cylinder internal construction as used by one manufacturer is shown in Fig. 18-11.

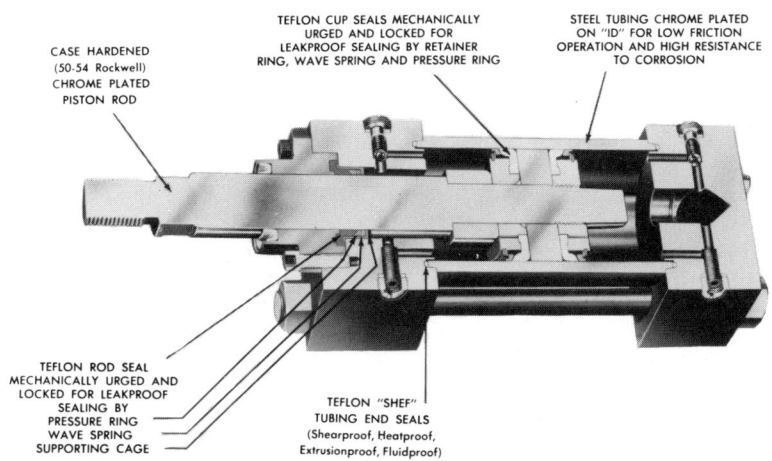

FIG. 18-11. Internal construction of a high temperature air cylinder. (*Courtesy* Miller Fluid Power, a Flick Reedy Subsidiary Corp.)

18-4 CYLINDER MOUNTING

Cylinders can be obtained with different types of standard mountings. These allow for rigid mounting of the cylinder or hinged mounting if the cylinder is required to rotate about a pivot during its actuation. Some of these mounts are described next.

Rabbet Mount

This mount (Fig. 18-12) is used for mounting on panels, bulkheads, and brackets. It mounts on the tapped holes in the front and rear face.

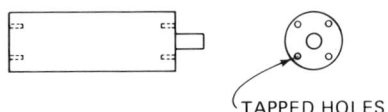

FIG. 18-12.

Flange Mount

The flange (Fig. 18-13) is designed to withstand high loadings. It may be obtained on the front as shown or on the rear of the cylinder.

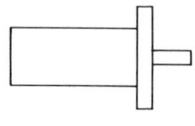

FIG. 18-13.

Foot Mount

This mounting (Fig. 18-14) is used for flat surfaces. It is a convenient and easy mounting method.

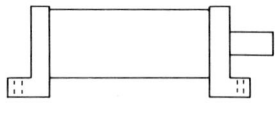

FIG. 18-14.

Centerline Mount

This mount (Fig. 18-15) has the advantage of keeping the reaction to the load in line with the centerline of the cylinder, eliminating any bending forces.

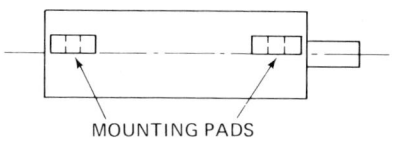

FIG. 18-15.

Clevis Mount

This is a very useful mount (Fig. 18-16) where the cylinder is required to pivot about a point at the rear.

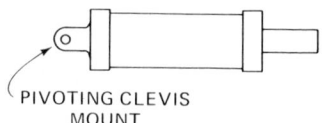

FIG. 18-16.

Trunnion Mount

This mount (Fig. 18-17) is also used where the cylinder is required to pivot about an axis.

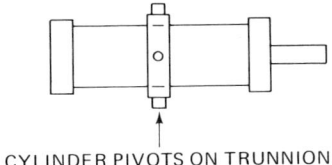

CYLINDER PIVOTS ON TRUNNION

FIG. 18-17.

In mounting a cylinder on a rigid mount, side loading such as that shown in Fig. 18-18 is to be avoided. The side loading tends to cause binding of the rod and results in excessive wear of the rod and cylinder.

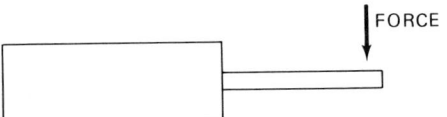

FIG. 18-18. Side load on a cylinder.

18-5 CUSHIONED CYLINDERS

Cushioned cylinders have been developed to slow down the piston as it reaches the end of its travel. This reduces the shock loading which occurs when the piston bottoms out at the end of the stroke. With high velocity the shock loading can result in excessive stresses in the cylinder.

The design of the cushioned cylinder is shown in Fig. 18-19. As the piston moves to the end of the stroke, the cushion nose enters the cushion chamber, slowing down the rate of flow of the fluid leaving the cylinder by forcing it through the small orifice shown. Cushioning on hydraulic cylinders is more effective than on air, because of the compressibility of air.

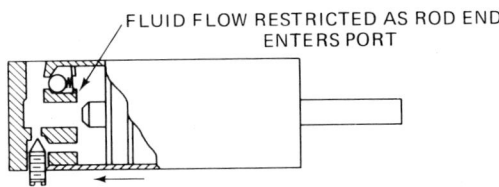

FIG. 18-19. Cushioned cylinder.

18-6   UNUSUAL CYLINDER APPLICATIONS

Two unusual ways of using cylinders are shown in Figs. 18-20 and 18-21. Figure 18-20 shows a cylinder with combination air–oil operation. The piston provides the driving force behind the rod. The oil is bled out of the system by the metering valve. This provides more stable operation of the rod and control over the feed rate. The disadvantage of the lack of speed control associated with air is offset somewhat with this system.

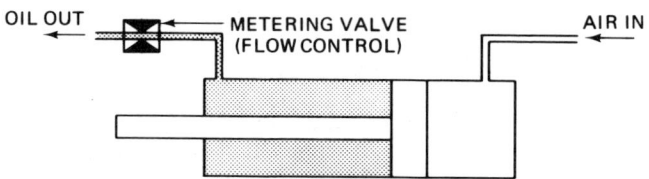

FIG. 18-20.   Combination air-oil operation to stabilize feed rate.

In Fig. 18-21 a cylinder is used as a shock absorber. This system requires that the fluid be noncompressible. A compressible fluid would act as a spring and rebound of the rod would occur. The fluid is metered through the orifice as the piston is forced back in the cylinder, slowing the rod and damping out the forces.

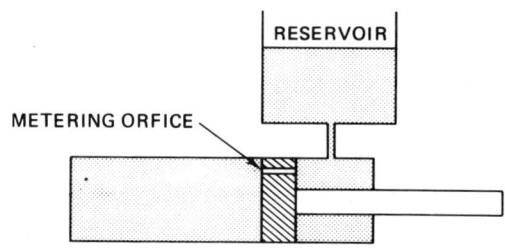

FIG. 18-21.   Hydraulic cylinder functioning as a shock absorber.

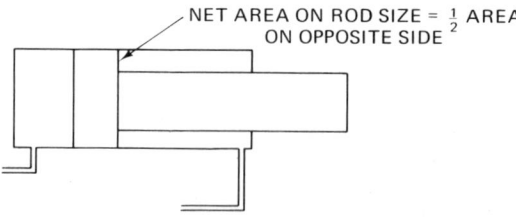

FIG. 18-22.   Hydraulic cylinder with return stroke velocity twice the velocity of extension stroke.

QUESTIONS AND PROBLEMS

A hydraulic cylinder that uses a large rod to provide a rapid return stroke is shown in Fig. 18-22. The net area on the rod side of the piston is one half the area on the other side. Since this is a hydraulic cylinder, the fluid is incompressible. The equation of continuity, $Q = Av$, then applies. If Q is constant on both strokes, then the velocity on the return stroke is twice the velocity on the extending stroke. This is true because the net area is one half the piston area.

QUESTIONS AND PROBLEMS

18-1.—Calculate the force exerted by the rod of the cylinder in Fig. 18-23 when the rod is extended and when it is returned.

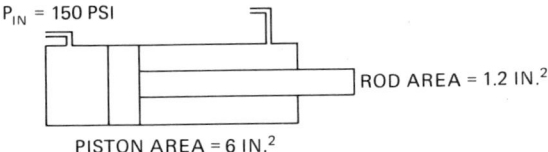

FIG. 18-23.

18-2.—Name four primary cylinder types as classified by cylinder function.

18-3.—What type of cylinder should be selected for the application in Fig. 18-24? Only 1 ft of vertical space is available for mounting the cylinder. One and one-half feet is required to complete contact with the die punch, then an additional 3-inch travel to complete the operation.

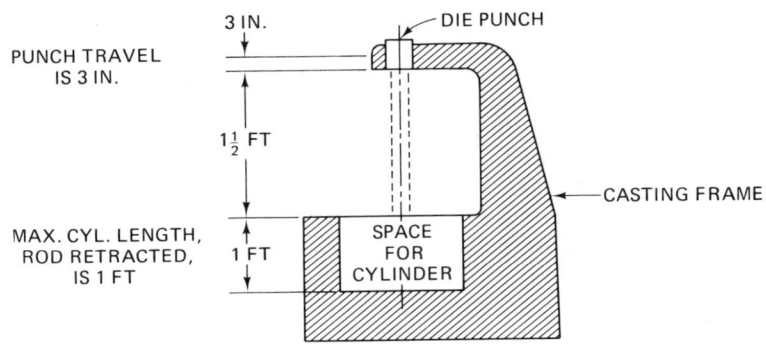

FIG. 18-24.

18-4.—What type of mounting should be selected for the application in Fig. 18-25?

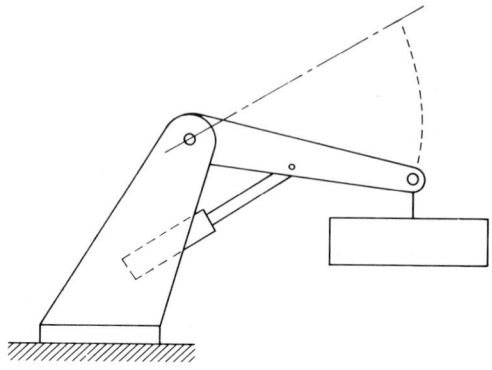

FIG. 18-25.

18-5.—Cylinders are frequently _____, to prevent shock loads at the end of a stroke.

chapter nineteen

FLUID POWER COMPONENTS—VALVES

19-1 INTRODUCTION

If the actuator does the work of the fluid power system, the *valve* might be likened to the operator who starts, stops, and controls the actuator. The valve does this by controlling flow quantity and pressure. With this in mind, a definition of a valve may be:

Valve: A mechanical device for controlling fluid flow quantity and/or pressure.

The control of flow quantity covers the full range from zero flow, with the valve shut, to the maximum flow when the valve is open. The control of pressure usually requires a special type of valve designed specifically for that purpose.

19-2 DIRECTIONAL CONTROL VALVES

Directional control valve is the general term used to describe those valves which control fluid flow quantity. The directional control grouping of valves is subdivided further into grouping by the number of ports or connecting lines in the valve. These openings are termed *ways*. The two-way valve has two ports, one for the fluid entry and the second for the fluid exhaust. The manually operated globe valve commonly used on the kitchen faucet is an example of a two-way valve. Two-way, three-way, and four-way directional control valve functioning may be visualized by examining Fig. 19-1. Directional control valves may be operated manually, electrically, by hydraulics, or by pneumatics.

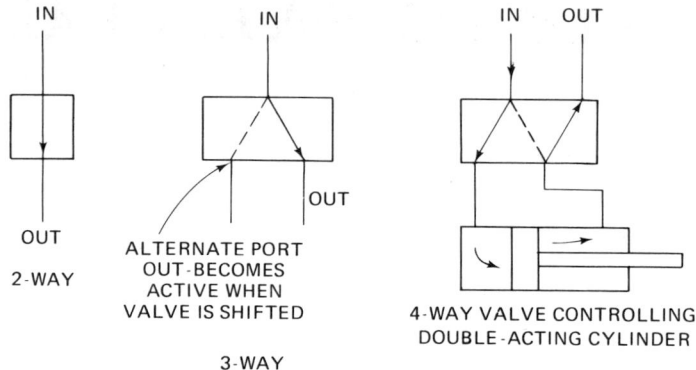

FIG. 19-1. Directional control valves.

Two common types of two-way valves are the *globe valve* and the *gate valve*, shown in Figs. 19-2 and 19-3, respectively. Both are manually operated. The gate valve is the more efficient of the two and has less pressure loss, as was pointed out in Table 15-1. Both types of valves are used for oil, water, and gas.

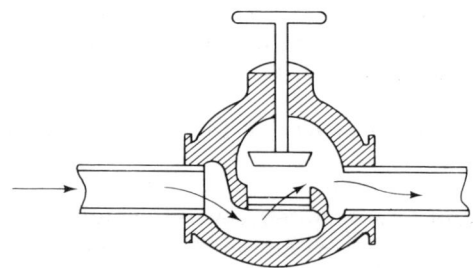

FIG. 19-2. Globe valve.

The *plug valve*, shown in Fig. 19-4 is an inexpensive, low-pressure valve. It has low pressure loss because of its full flow when fully open and only requires a quarter turn to open. It is not recommended for high pressure because when closed the high loading on the valve makes it difficult to open.

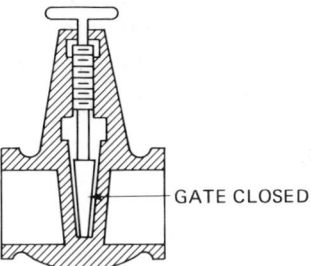

FIG. 19-3. Gate valve.

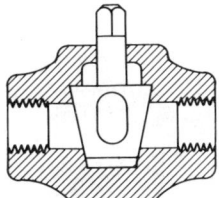

FIG. 19-4. Plug valve.

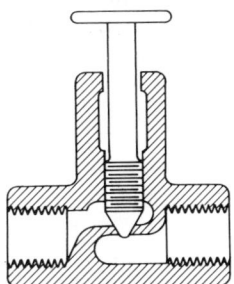

FIG. 19-5. Needle valve.

The *needle valve* in Fig. 19-5 is used primarily to throttle the flow of fluid in a line. It is not designed to be used as a shut-off valve, as are the globe, gate, and plug valves. The needle valve is also manual in operation.

19-3 SPOOL-TYPE DIRECTIONAL CONTROL VALVES

The *spool-type directional control valves* are used in fluid power circuits because of the ease with which they are shifted to perform a switching function in the circuit. The ease in shifting, which diverts fluid from one flow path to another, is the result of the type of construction used in the valve.

Typical construction is shown in the cut-away view of the Vickers directional control valve in Fig. 19-6 and in the drawing in Fig. 19-7. Both of the valves in these figures are manually operated. The shifting of the spool as shown in Fig. 19-7 changes the flow of fluid.

FIG. 19-6. Vickers manually operated directional control valve. (*Courtesy* Vickers division, Sperry Rand Corp.)

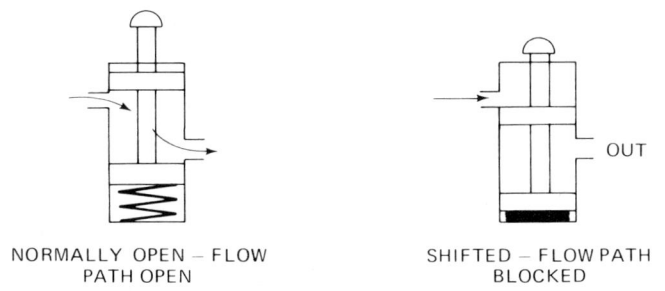

NORMALLY OPEN — FLOW PATH OPEN

SHIFTED — FLOW PATH BLOCKED

FIG. 19-7. Two-way directional control valve.

The three-way directional control valve is shown in Fig. 19-8. The valve shown is spring-loaded and manually operated, but other types of operation can be obtained. The flow may also be opposite to the direction shown.

The four-way valve is shown in Fig. 19-9. The valve is considered four-way, although there are five ports. Exhaust ports 1 and 2 perform the same function and are considered to be one port. In hydraulic valves these ports are usually connected inside the valve and brought out as one port. Air valves may use the two-port arrangement as shown.

More combinations of flow paths are possible with four-way valves.

Sect. 19-3 / SPOOL-TYPE DIRECTIONAL CONTROL VALVES

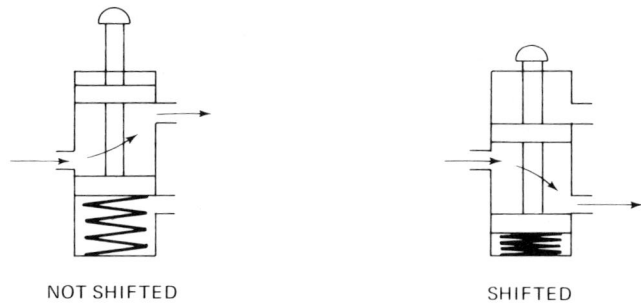

FIG. 19-8. Three-way directional control valve.

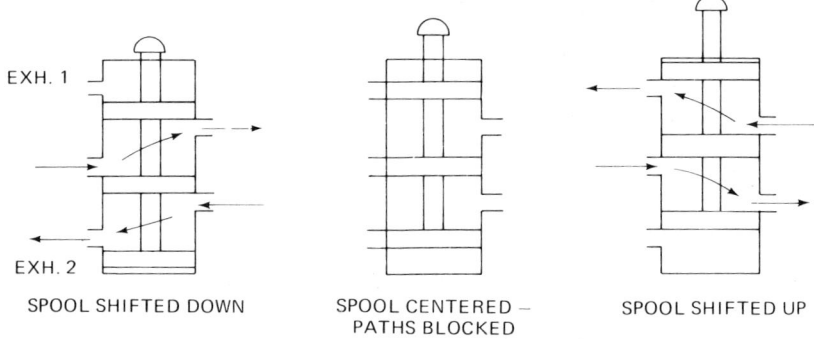

FIG. 19-9. Four-way, three-position directional control valve.

In the valve in Fig. 19-9 the spool may be in any one of three positions. If the spool is in the center position, the flow is blocked in all paths. Two-position valves are also available. In these the spool is in only one of two available positions. The two-way and three-way valves discussed previously are two-position valves. A two-position, four-way valve is shown in Fig. 19-10.

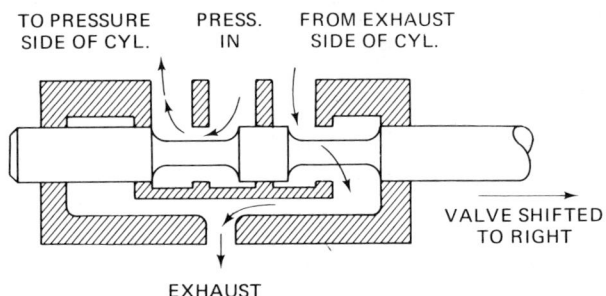

FIG. 19-10. Four-way, two-position valve.

Three types of the four-way, three-position valves are the closed-center, open-center, and tandem-center spool. The function of the closed-center spool is that as shown in Fig. 19-9, where the flow is blocked. Flow paths obtained in the other types are shown in Figs. 19-11 and 19-12.

Cutaway views of solenoid-operated valves are shown in Figs. 19-13 and 19-14. The operation of the valve in Fig. 19-13 is by direct solenoid action. The valve in Fig. 19-14 is pilot-operated by the solenoids and the main spool is shifted by hydraulic action.

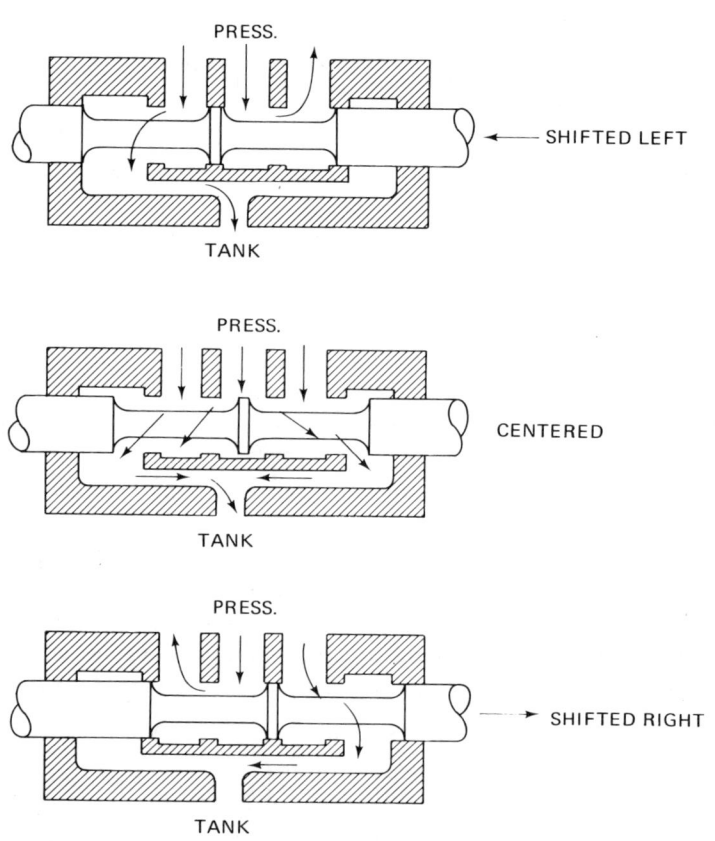

FIG. 19-11.   Open center spool—four-way valve.

19-4   APPLICATION OF DIRECTIONAL CONTROL VALVES

The applications of directional control valves to fluid power system are so numerous and diverse that only a few can be mentioned here. One of the

Sect. 19-4 / APPLICATION OF DIRECTIONAL CONTROL VALVES

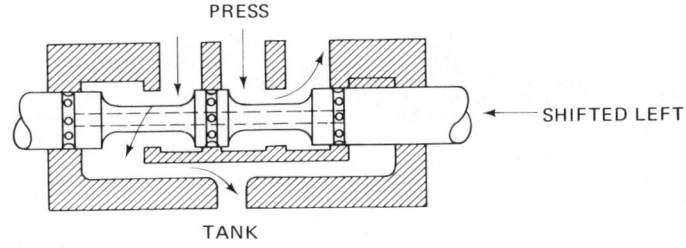

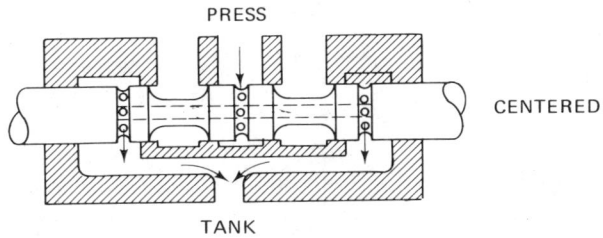

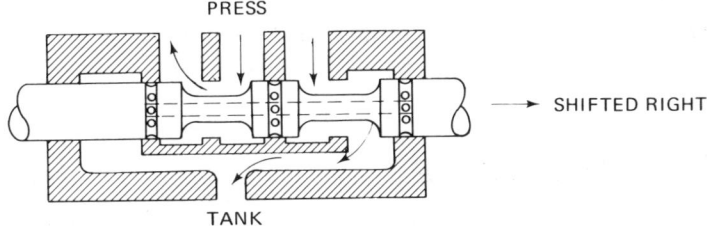

FIG. 19-12. Tandem-center spool four-way valve.

FIG. 19-13. Direct Solenoid operated valve. (*Courtesy* Miller Fluid Power, a Flick-Reedy subsidiary.)

most important is the use of directional control valves to operate cylinders and other types of actuators. The applications described in Fig. 19-15 are for three- and four-way valves.

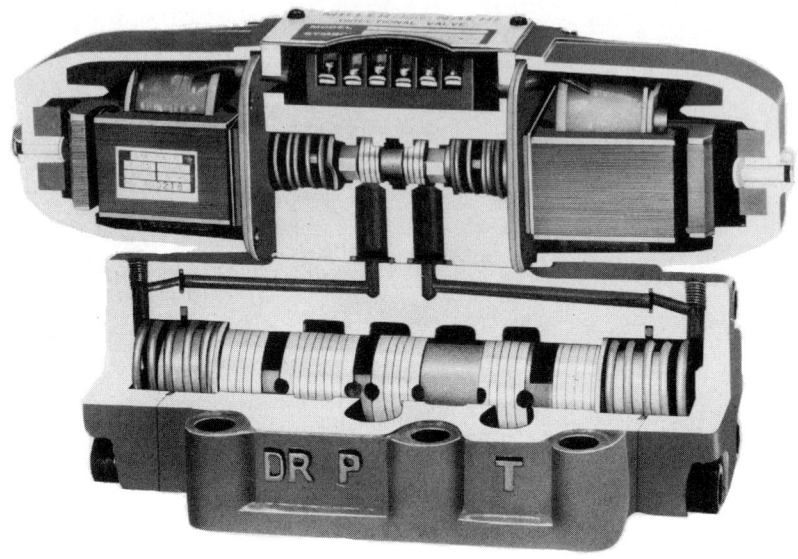

FIG. 19-14. Solenoid-piloted directional control valve. (*Courtesy* Miller Fluid Power, a Flick-Reedy subsidiary.)

19-5 CHECK VALVE

A special type of directional control valve is the *check valve*. The check valve permits flows in one direction but not in the opposite direction. It is a two-way valve used in all types of systems ranging from the municipal water supply system to the fluid power system. Typical construction of some types of check valves are shown in Fig. 19-16.

19-6 PRESSURE CONTROL VALVES

The pressure regulator shown in Fig. 17-8 is a type of *pressure control valve*. Fluid power systems require pressure control to assure proper functioning of the components as well as to provide safety against excessive pressure buildup in the system.

The pressure relief valve is used in all hydraulic systems and other systems in which high pressures may occur. In controlling the maximum pressure in the system, it functions as a safety device. This is particularly important in a hydraulic system where a positive displacement pump is used. The term "positive displacement" would apply to a piston pump, since movement of the piston has to be reflected in a displacement of the

Sect. 19-6 / PRESSURE CONTROL VALVES

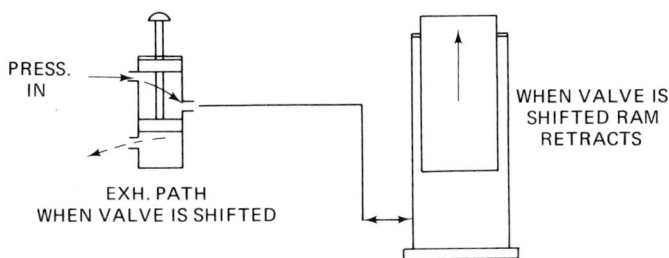

(a) 3-WAY VALVE USED TO RAISE HYDRAULIC RAM

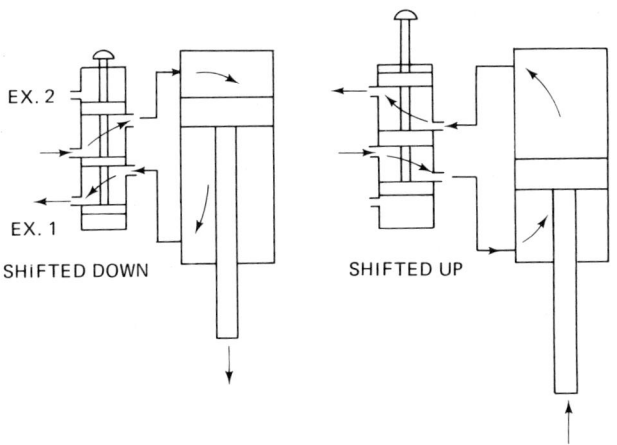

(b) 4-WAY VALVE CONTROLLING DOUBLE-ACTING CYLINDER

FIG. 19-15.   Directional control valve applications.

incompressible fluid in the hydraulic system. If the fluid movement is blocked in any way, excessive pressures build up very rapidly in the system.

Relief valves generally are designed to work against spring loading as shown in Fig. 19-17. The pressure at which the valve opens can be preset into the valve. The working pressure of the system is not generally determined by the relief valve. Only maximum pressure is determined by it.

The correct location of a pressure relief valve is shown in Fig. 19-18. The valve is placed on the discharge side of the pump before any other valves. A similar location in an air system would be just after the air compressor on the high-pressure side.

Relief valves are available which perform more complicated circuit functions. The Vickers multi-pressure relief valve shown in Fig. 19-19 is

194 FLUID POWER COMPONENTS—VALVES / Chap. 19

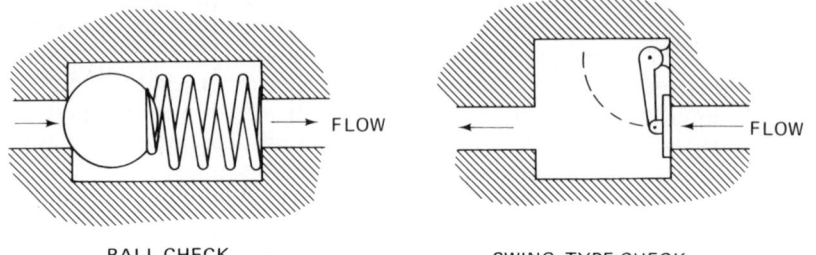

BALL CHECK   SWING-TYPE CHECK

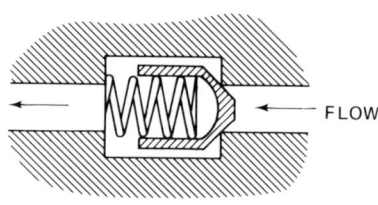

CONE TYPE CHECK

CHECK VALVES

FIG. 19-16. Check valves.

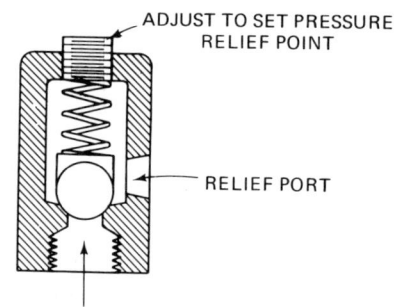

FIG. 19-17. Relief valve.

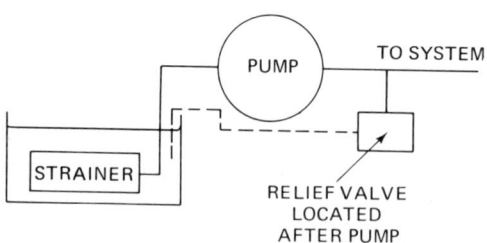

FIG. 19-18. Relief valve location.

Sect. 19-6 / PRESSURE CONTROL VALVES

FIG. 19-19. Vickers multi-pressure relief valve. (*Courtesy* Vickers division, Sperry Rand Corp.)

one example. Construction is similar to a standard four-way directional valve, but the valve functions as a relief valve at three different pressure settings. Three pressure ranges are available: (1) 125 to 1000 psi, (2) 500 to 2000 psi, and (3) 1500 to 3000 psi.

Unloading relief valves are designed to unload excessive flow from a hydraulic pump by diverting the flow back into the reservoir when the

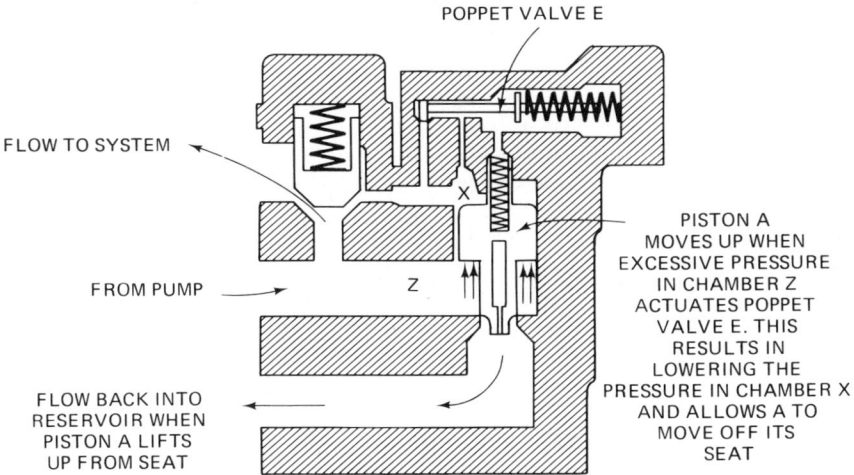

FIG. 19-20. Unloading relief valve.

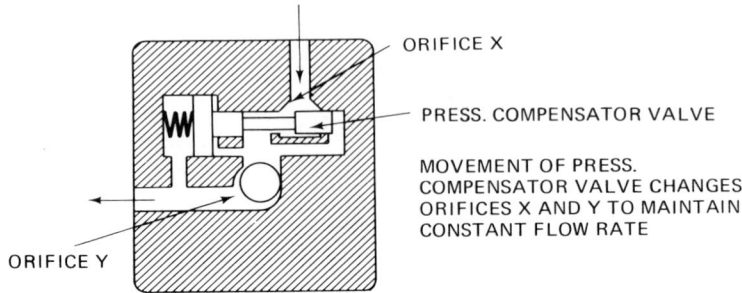

FIG. 19-21. Flow control valve.

pressure reaches a certain preset value. The functioning of this type of valve is shown in Fig. 19-20.

19-7 FLOW CONTROL VALVES

Flow control valves are designed to provide control of the fluid quantity in a hydraulic circuit. They are adjustable but once set they maintain a constant flow rate regardless of the change of load. The constant flow rate is accomplished by automatic compensation of pressure changes within the valve. Functioning of this type of valve is explained in Fig. 19-21.

19-8 PILOT OPERATION OF VALVES

The shifting of the spool in a directional control valve can be done by hydraulic means, pneumatics, electrical power, and mechanical power.

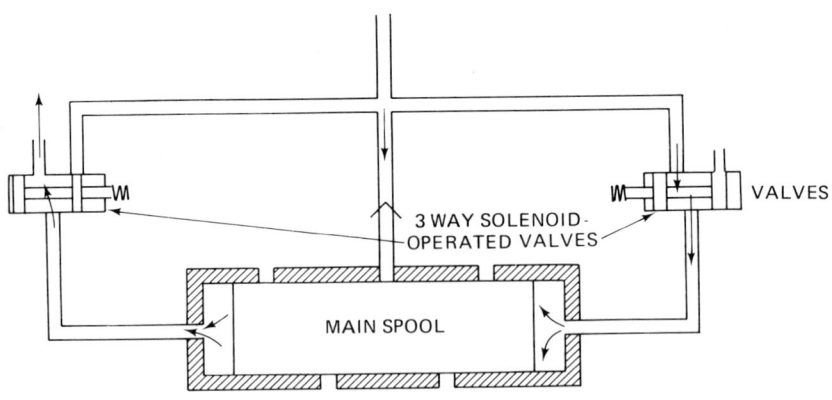

FIG. 19-22. Pilot-operated valve.

Smaller valves may have the spool shifted directly by electrical solenoids. The *solenoid* is a type of electromagnet that can be adapted to perform the push–pull type of movement necessary to shift the spool. If large valves and high pressures are involved, the solenoid becomes unsatisfactory as a result of high current requirements, heating, and short life. For this situation the spool may be shifted by means of pilot operation.

The principle of pilot operation is shown in Fig. 19-22. The fluid that operates the spool may be the system fluid or a separate supply, since they are separated in the valve. The pilot valves operate the main valve. If the pilot valves are operated by direct solenoid action, the main valve operation is described as solenoid-controlled, pilot-operated.

QUESTIONS AND PROBLEMS

19-1.—Why is the gate valve considered to be more efficient than the globe valve?

19-2.—Select a directional control valve for a reversible hydraulic motor that will reverse the motor and stop the motor. That is, by operating one valve, the motor can be made to run (a) clockwise, (b) counterclockwise, and (c) stop rotation.

19-3.—Select a valve to be installed where shown in Fig. 19-23 to allow the tank to be vented to atmosphere and then evacuated.

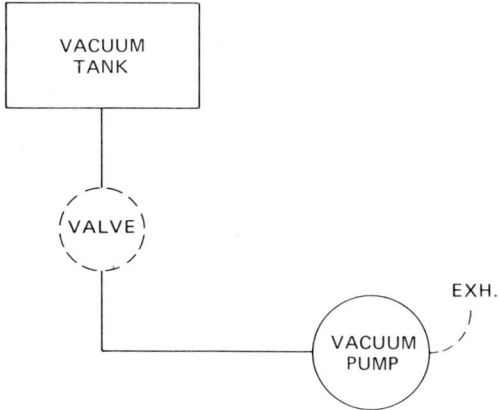

FIG. 19-23.

19-4.—A hydraulic motor requires 5 gpm of hydraulic fluid to operate at rated torque and speed. What type of valve should be selected to ensure that the motor has the required flow?

19-5.—The pump in Fig. 19-24 pumps water 6 ft up into the tank. What type of valve should be installed to prevent return flow in the line automatically if the pump is shut off?

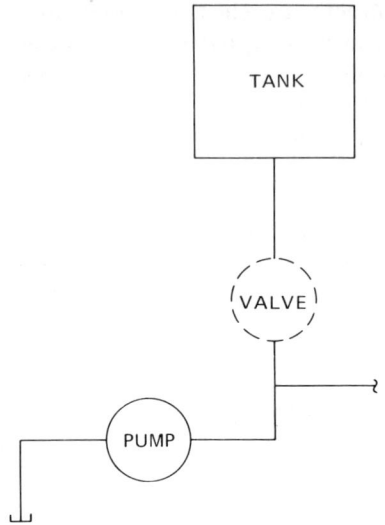

FIG. 19-24.

19-6.—For large directional control valves subjected to high pressure, direct solenoid operation of the valve is generally unsatisfactory because of _____. (*Select the correct answer.*)

a—Too slow solenoid operation.

b—The requirement for low voltage input current to the solenoid.

c—Heating and high current requirement of the solenoid.

d—Difficulty in mechanically coupling the spool to the solenoid.

19-7.—For the large valve in Problem 19-6, what method of operation may be used to overcome the disadvantage of direct solenoid operation?

19-8.—The plug valve is not suitable for high pressure because _____. (*Select the correct answer.*)

a—It has excessive pressure loss.

b—It leaks.

c—It requires an excessive number of turns to open.

d—High loading from the pressure makes it difficult to turn.

QUESTIONS AND PROBLEMS 199

19-9.—A needle valve is used for _____. (*Select the correct answer.*)

a—Throttling.

b—Full open and shut off for high-pressure lines.

c—Controlling three-way directional control valves.

d—Safety to prevent excessive pressure buildup.

19-10.—A pressure relief valve is required for the system in Fig. 19-25. Where should it be located? (*Select the correct answer.*)

a—At A, between valves and pump.

b—At B, between valves and each actuator.

c—At C, in return line to reservoir.

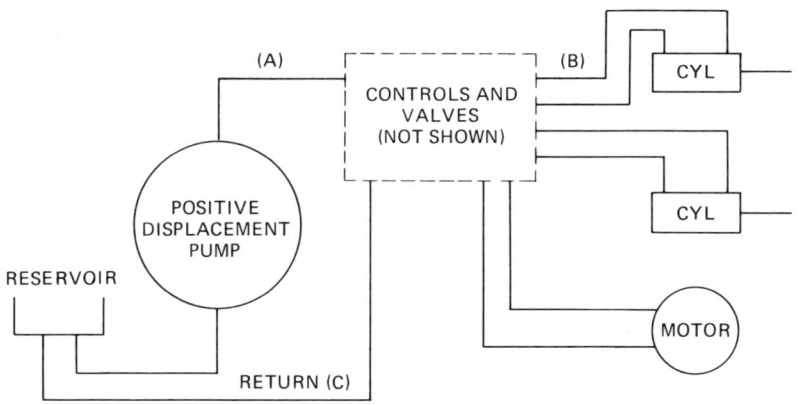

FIG. 19-25.

chapter twenty
PIPING AND TUBING— INTENSIFIERS

20-1 INTRODUCTION

The choice of type of fluid conductor to use in a fluid power system is determined by various factors. In the end, the choice may be to use several types of conductors in the same system.

The factors below are common ones to consider in selecting the type of conductor:

1 / Cost.
2 / Pressure requirements.
3 / Corrosion resistance to the system fluid and also to the environment in which the pipe or tubing is located.
4 / Resistance to abuse and abrasion.
5 / If relative movement of the system components occurs, flexible tubing is required.
6 / Shock loading, which occurs with pressure surges.
7 / Minimum and maximum temperatures.

Commercial steel pipe, sometimes called black steel pipe, is the cheapest material to use. Built-up industrial hose is used for severe operating conditions of high pressure, shock, and where abrasion is high.

20-2 PIPE

Commercial steel pipe is furnished in standard sizes according to an ANSI (American National Standards Institute) standard. The standard classifies pipe size by a schedule number. Prior to the use of schedule numbers, steel pipe was classified as standard, extra heavy, and double extra heavy. Schedule numbers 40 and 80 correspond to standard and extra heavy. Schedule 160 falls between the extra heavy and double extra heavy classifications. The outside diameter of the pipe remains the same for all schedule numbers. The wall thickness increases as the schedule number increases. The pipe cross sections in Fig. 20-1, although not to scale, give a picture of the relative size variations. Table VI in the Appendix shows dimensions of Schedule 40 and Schedule 80 pipe.

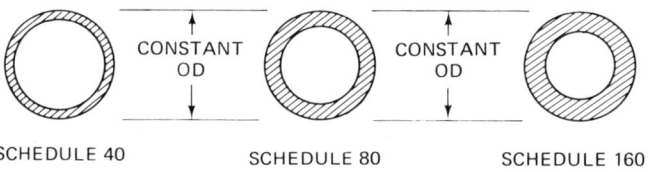

FIG. 20-1. Relative cross-sectional areas of steel pipe.

A primary consideration in determining which pipe schedule to use is the internal pressure of the system. An approximate equation that may be used is the following one:

$$\text{schedule number} = 1000 \frac{P}{S} \qquad (20\text{-}1)$$

where P = internal pressure, psig
       S = allowable stress value of the steel

The value of S depends upon the grade of steel used. If carbon steel pipe with an allowable stress of 6750 psi is used, Schedule 40 pipe can be used for pressures up to 270 psig when Eq. (20-1) is used to calculate the pressure.

Steel pipe may be obtained in higher-strength alloy steels. These higher-strength alloy steels allow higher working pressures than the carbon steel noted above. Several of these are noted in Table 20-1.

Table 20-1   Allowable Stresses for Pipe

Material	Stress, S, at −20 to 650°F
Butt-welded A53 carbon steel	6,750
Carbon-molybdenum steel, A204	14,600
A312 stainless steel	12,650

*Specification numbers of these steels are from the ASTM Standard Code for Pressure Piping.

20-3   THREADS

Threads for pipe and fittings are either American National Standard NPT or NPTF threads. The NPT pipe thread is that which is commonly used for plumbing systems and requires a thread sealer to prevent leakage. The NPTF thread (Dryseal) seals by interference of the threads and does not require a sealer unless it is reused. SAE (Society of Automotive Engineers) standards require the NPTF thread.

20-4   TUBING

Both rigid and flexible tubing can be obtained for use in hydraulic and pneumatic lines. Material for the flexible tubing is plastic.

Tubing is available in the following materials:

1 / Soft, low carbon steel: can be bent (unlike pipe).
2 / Aluminum: approved only for low pressure.
3 / Copper: should not be used with hydraulic systems because it acts as an oil oxidizer and work hardens when bent or flared. This makes it susceptible to fatigue failure. Copper is satisfactory for low-pressure, stationary applications in pneumatics.
4 / Plastic: available in polyvinyl chloride (PVC), polyethylene, nylon, and tetrafluoroethylene (TFE). Polyvinyl chloride can be used for pressures to 125 psi and temperatures to 100° F continuously. Nylon may be used to 250 psi and TFE up to 1000 psi.

Table 20-2 may be used for the selection of steel tubing to withstand the system pressure. This table lists JIC (Joint Industry Commission) specifications for selecting tubing by wall thickness for two pressure ranges, 0 to 1000 psi and 1000 to 2500 psi.

Table 20-2   Pressure vs. Wall Thickness for Steel Tubing

Tubing Outside Diameter (in.)	0–1000 psi Wall Thickness (in.)	1000–2500 psi Wall Thickness (in.)
0.250	0.035	—
0.312	0.035	—
0.375	0.035	0.058
0.500	0.042	—
0.625	0.049	0.095
0.750	—	0.120
0.875	0.072	—
1.000	—	0.148
1.250	0.109	0.180
1.500	0.120	0.220

20-5   HOSE

Built-up hose is available for heavy-duty applications where relative movement between the two connected components is required. A very common example is the hose connection to the front wheels of an automobile, where movement of the wheel vertically and about a nearly vertical axis when turning are both present.

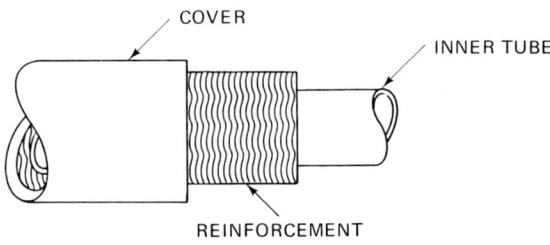

FIG. 20-2.   Typical hose construction.

Hose is furnished in inside-diameter ranges from $\frac{3}{16}$ to 3 inches. With metal wire braid reinforcement working pressures of up to 1500 psi can be used. With ordinary fabric or plastic braid, pressures up to 500 psi can be used. Typical hose construction can be seen in Fig. 20-2. The inner tubing or core may be of several different materials but should be selected so as to be compatible with the fluid.

20-6 FITTINGS FOR TUBING AND HOSE

Some of the various types of fittings and connectors available for tubing and hoses are shown in Fig. 20-3.

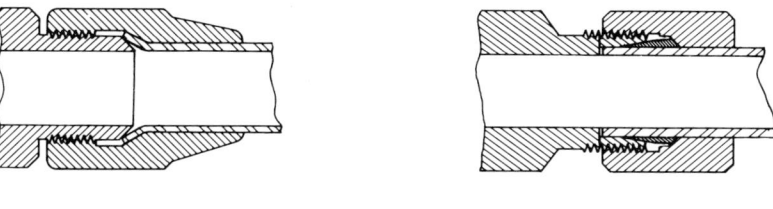

FLARE FITTING   FERRULE FITTING

QUICK DISCONNECT
BUTT SEAL AND BALL TYPE

FIG. 20-3. Fittings and connectors.

20-7 INTENSIFIERS

Intensifiers, sometimes called *boosters*, are used to increase the pressure in a system. In electronics the same function would be called *amplification*. The intensifier takes a lower pressure from a system and increases it to a higher pressure. The higher-pressure side of the intensifier may use fluid from the same system or it may use fluid from another system.

The operation of the intensifier can be seen by examining Fig. 20-4. The principle rests on Pascal's law. Using the equation $F = PA$, the force F exerted on the 2-in. piston is the same as the force the 4-in. piston exerts on the rod. Then

$$F = P_1 A_1 = P_2 A_2 \quad \text{and} \quad P_1 A_1 = P_2 A_2$$

Solving for P_2,

$$P_2 = \frac{P_1 A_1}{A_2}$$

Very high pressures are obtainable with intensifiers.

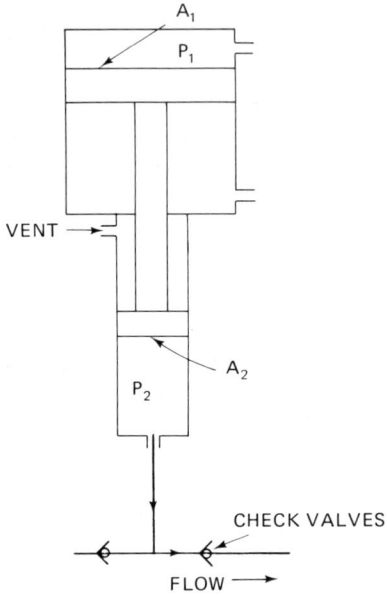

FIG. 20-4.   Pressure intensifier.

QUESTIONS AND PROBLEMS

20-1.—What pipe schedule should be used for A204 chrome molybdenum steel if the maximum system pressure is 1000 psi?

20-2.—What thread system is recommended by the Society of Automotive Engineers for pipe?

20-3.—Name three plastics that are extensively used for tubing.

20-4.—What is the disadvantage of copper when used in a hydraulic system?

chapter twenty-one

PUMPS

21-1 INTRODUCTION

Two broad classifications of pump types exist. One is the positive-displacement type and the other is the hydrodynamic type. The positive-displacement type is more important in fluid power.

The functioning of both of these types can be seen by examining Fig. 21-1. The positive displacement pump transfers a fixed volume of incompressible fluid through it. The piston pump in Fig. 21-1(a) draws in fluid into the cylinder as the piston is raised on the suction stroke. The two check valves shown will allow flow to the right but prevent it in the opposite direction. On the downstroke of the piston, fluid is forced out and to the right. The check valve on the left prevents fluid flow back into the reservoir. The fixed volume of fluid that had been in the cylinder has now been transferred out of the cylinder, thus giving rise to the term "positive displacement."

Now examine the centrifugal pump in Fig. 21-1(b). This pump is of the hydrodynamic type. Its operation depends on the conversion of the incoming fluid static pressure head to a high-velocity head. The rotating impeller increases the velocity of the fluid particles at the periphery of the impeller, increasing the kinetic energy of the fluid. There is slippage be-

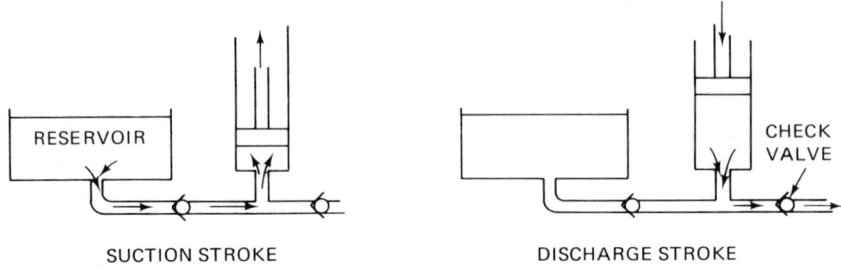
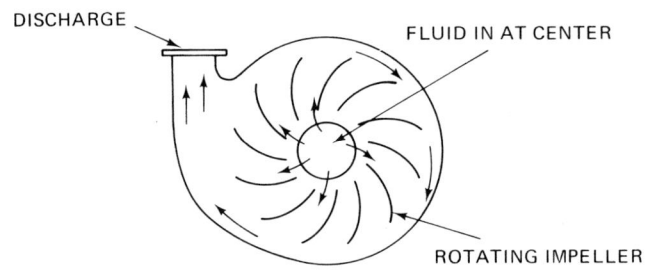

FIG. 21-1. Pump types.

tween the impeller and the fluid near the outer periphery of the impeller. Because of this the pump does not qualify as a positive-displacement pump. Since positive-displacement pumps are used almost exclusively in the fluid power field, only this type is discussed here.

21-2   CAVITATION IN PUMPS

A potentially destructive condition exists in a pump when the inlet absolute drops to a point equal to the vapor pressure of the liquid being pumped. This condition is called *cavitation*. It may occur under conditions as shown in Fig. 21-2. Here a pump is handling gasoline with a vapor pressure of 4.42 psia at 70° F. The pump is operating under a suction head on the inlet side. If the distance X becomes great enough to reduce the pressure at the pump inlet to 4.42 psia, the liquid will start to vaporize and form gaseous bubbles. This phenomenon is cavitation. It can cause vibration of the pump rotating members, erosion of the metal, and failure of pump components. Air leaking into the suction from other sources generally can have the same effect.

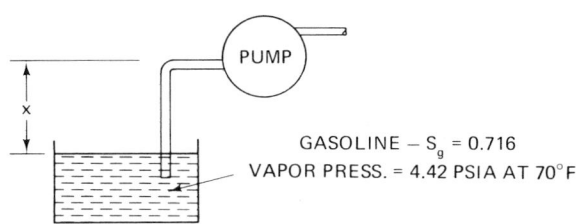

FIG. 21-2.   Pump cavitation.

Proper design of pump suction is the best assurance against cavitation.

21-3   GEAR PUMPS

One of the most popular types of positive-displacement pump is the gear pump. A cutaway section of a Vickers G-20 gear pump is shown in Fig. 21-3. Functioning of this type of pump is explained in Fig. 21-4. The fluid is transferred through the gear pump in the space between the gear teeth.

Backflow is prevented by the sealing of the teeth as they contact at the center of the pump. This is accomplished by having one gear as the driver and the other gear as the driven member. Contact between gear teeth is assured by this method.

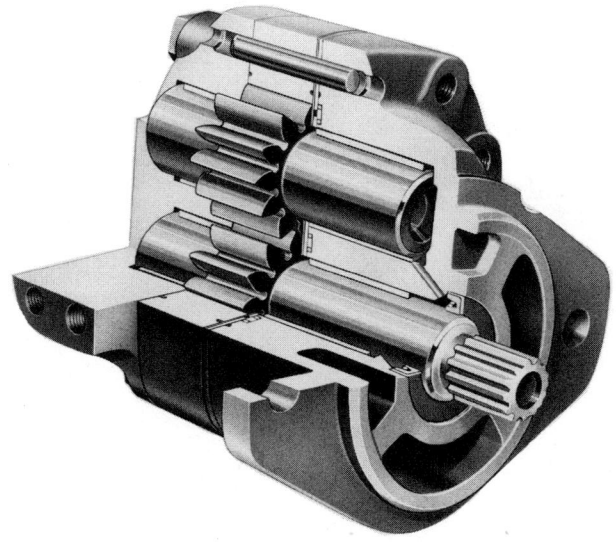

FIG. 21-3.   Vickers G-20 gear pump. (*Courtesy* Vickers division, Sperry Rand Corp.)

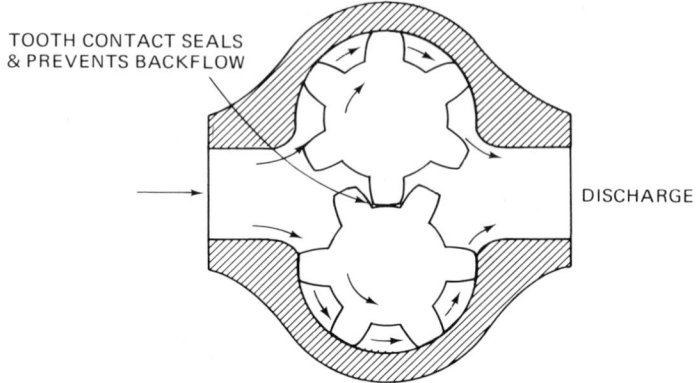

FIG. 21-4. Functioning of gear pump.

Also available is the internal gear pump. The noise level of the internal gear pump is somewhat lower than the external gear, but sensitivity to dirt and other material in the fluid is greater.

21-4 VANE PUMPS

Figure 21-5 illustrates a cutaway section of a Vickers 5045v double-vane hydraulic pump. This is really two pumps located side by side, a not uncommon arrangement in hydraulic pumps. This type of construction pro-

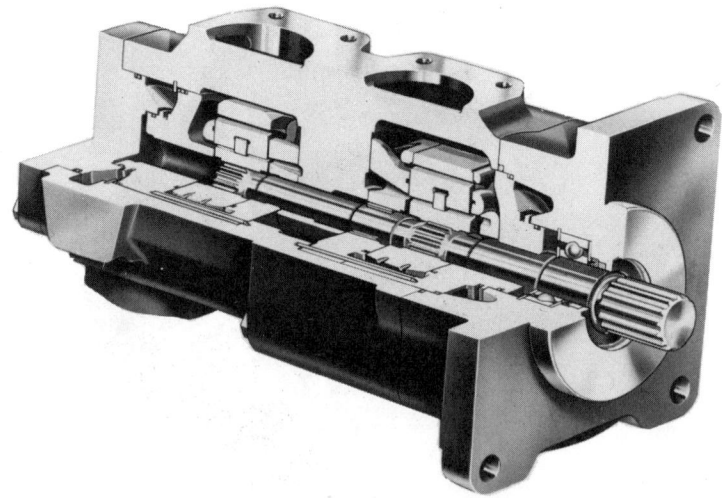

FIG. 21-5. Vickers 5045V Double vane hydraulic pump. (*Courtesy* Vickers division, Sperry Rand Corp.)

vides more capacity than a single unit, yet it functions as a single unit. In this particular pump the small pump element is rated up to 2500 psi and the large element up to 2000 psi. Flow capacity goes up to 109 gpm for the large element.

The principle of operation of the unbalanced vane pump is shown in Fig. 21-6. The vanes in the rotor can move in the slots but are held out against the outer casing by springs, by centrifugal force, or fluid pressure. The rotor is eccentrically mounted in the housing. As it rotates, the volume between the rotor and housing increases. Since the vanes seal this volume off, the result is a reduced pressure and vacuum in this volume. As rotation continues, the volume decreases and pressure builds on the discharge side of the pump.

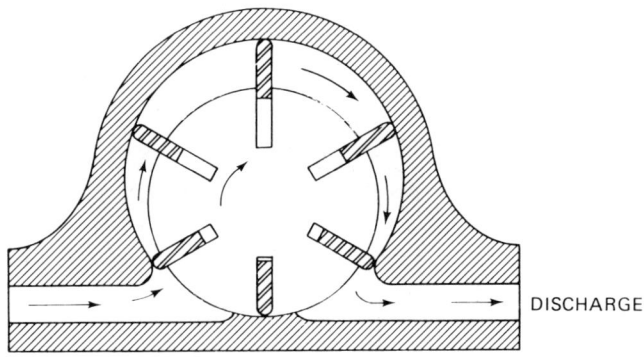

FIG. 21-6. Unbalanced vane pump.

The unbalanced vane pump causes fairly heavy radial loads on the bearings because of the low-pressure area on one side. There are balanced vane-pump designs which eliminate this objection. This is done by providing two low-pressure areas 180° opposite each other, thereby canceling the side forces.

21-5 PISTON PUMPS

Two types of piston pumps are the axial piston and the rotary piston pump. In both these types the pumping action comes from the reciprocation of the pistons in the cylinders. The mechanisms required to accomplish this are complicated. Simplified functional diagrams of these mechanisms are shown and explained in Figs. 21-7 and 21-8. Note the completely different methods used to achieve reciprocations of the pistons in the cylinders in the two designs.

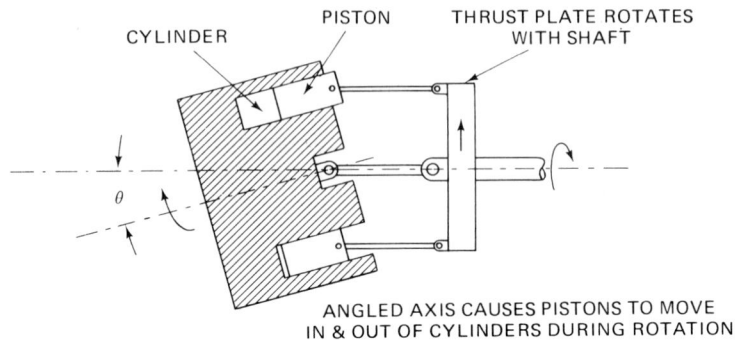

FIG. 21-7. Axial piston pump.

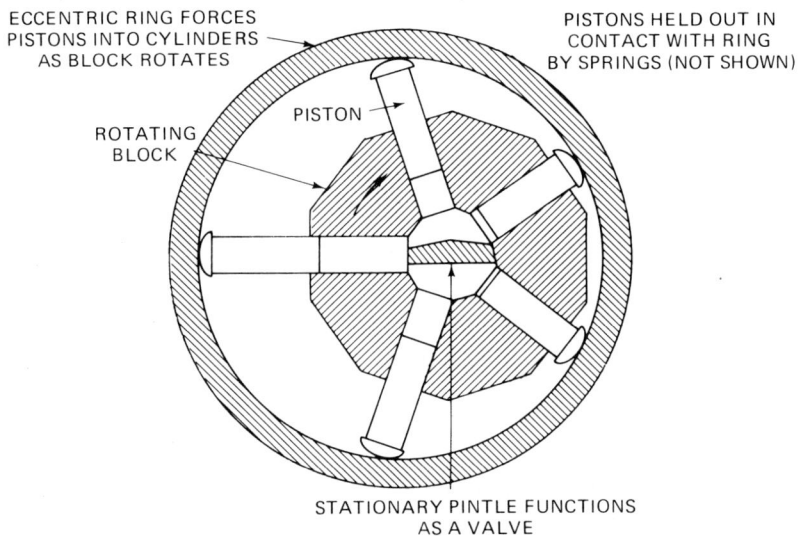

FIG. 21-8. Rotary piston pump.

21-6 CHARACTERISTICS OF DIFFERENT TYPES OF PUMPS

The three classes of positive-displacement pumps described here represent only the major ones used in the fluid power field. Diaphragm pumps and screw pumps are some of the other types which are not covered here.

It is frequently useful to have a knowledge of the characteristics of the three classes of pumps. Table 21-1 summarizes some of these characteristics. Comparative performance curves of the three types are shown in Fig. 21-9. These curves are approximate only and should be used only in estimating relative characteristics of the pump types.

Table 21-1 Some Characteristics of Positive-Displacement Pumps

Pump Type	Range of Pressures (psi)	Delivery Range (gpm)	Size Range (lb)
Gear	100–2500	0.2–300	7–530
Vane	1000–2000	1.5–77	11–240
Piston	300–5000	7.5–2300	20–43,000

(*Note:* All figures are approximate and may vary in certain cases.)

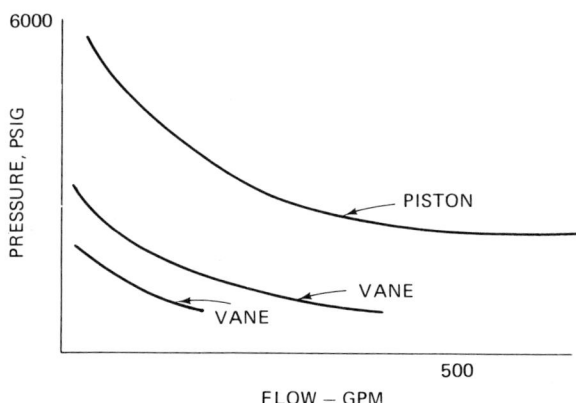

FIG. 21-9. Performance characteristics of piston, gear and vane power.

QUESTIONS AND PROBLEMS

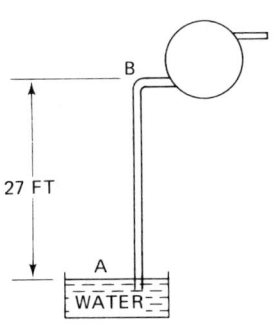

21-1.—The friction loss on the suction side of the pump in Fig. 21-10 from A to B is 2 psi. If the fluid is water and the suction height is 27 ft as shown, is cavitation likely to occur at the pump? Explain the method you used to arrive at your conclusion.

FIG. 21-10.

21-2.—What are the two major classes of pump types? Which one of these is used almost exclusively in fluid power?

21-3.—What type of pump should be selected for an output pressure of 3500 psi?

21-4.—Describe the operation of the rotary piston pump.

21-5.—What are some of the effects of cavitation?

21-6.—If the angle θ in Fig. 21-7 between the cylinder axis and the shaft of the axial piston pump is zero, what is the pump output?

21-7.—The balanced vane pump has the advantage of _____ as compared to the unbalanced vane pump. (*Select the correct answer.*)

a—Lower cost.

b—Higher speed.

c—Greater pressure.

d—Eliminating the heavy radial load on the bearing.

21-8.—Explain the term "positive displacement" as used in describing a hydraulic pump.

21-9.—In a gear pump, backflow at the center of the pump is prevented by maintaining positive tooth contact at this point. How is this accomplished?

chapter twenty-two

MOTORS AND HYDROSTATIC TRANSMISSIONS

22-1 INTRODUCTION

Motors are made for hydraulics and pneumatics and are therefore applied to both compressible and incompressible fluids. It is interesting to note a distinction here between motors and pumps. The pumps described in Chapter Twenty-one are for incompressible fluids. For the compressible fluid the compressor takes over the pump function.

The similarities and differences mentioned above become more interesting when it is discovered that the design and construction of many pumps and motors are almost the same. Indeed, in some cases the same unit can be used as a pump or as a motor.

22-2 MOTOR TORQUE AND POWER

From Fig. 12-1 in Chapter Twelve, a characteristic output of a motor was listed as *torque*. Torque when multiplied by the angular velocity of the output shaft yields the *power term* for the motor. This is explained in more detail in Fig. 22-1 and includes a basic definition of torque.

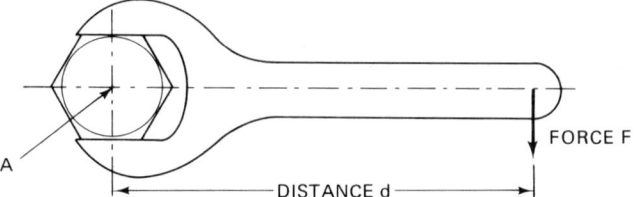

TORQUE = FORCE × PERPENDICULAR DISTANCE FROM FORCE F TO POINT A.
TORQUE = Fd

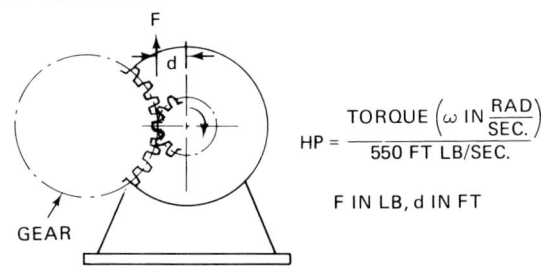

MOTOR ROTATING AT ANGULAR VELOCITY OF $\omega \frac{\text{RADIANS}}{\text{SEC.}}$

POWER = TORQUE × ω

$$HP = \frac{\text{TORQUE} \left(\omega \text{ IN } \frac{\text{RAD}}{\text{SEC.}}\right)}{550 \text{ FT LB/SEC.}}$$

F IN LB, d IN FT

FIG. 22-1. Torque and horsepower for a motor.

The displacement of a motor and also a pump is defined as the volume of the motor or pump chamber. For an incompressible fluid, the volume of fluid handled by the motor or pump in one revolution is equal to the volume of the chamber. Because of losses due to leakage and friction, the actual fluid volume is less than the theoretical volume. The ratio of actual to theoretical volume, multiplied by 100 percent, is the volumetric efficiency:

$$\text{volumetric efficiency} = \frac{\text{actual volume discharged}}{\text{theoretical volume}} \times 100\%$$

The torque of a hydraulic motor is proportional to the pressure and displacement:

$$\text{torque} = \frac{\Delta P \times \text{displacement/rev}}{2\pi} \qquad (22\text{-}1)$$

where ΔP is the pressure drop across the motor.

Torque characteristics for the air motor are different from those of the hydraulic motor. Theoretically, the torque of the air motor is a function of the pressure difference across the motor and a constant determined by the motor design. Motor speed therefore should have no effect on the torque

Sect. 22-2 / MOTOR TORQUE AND POWER

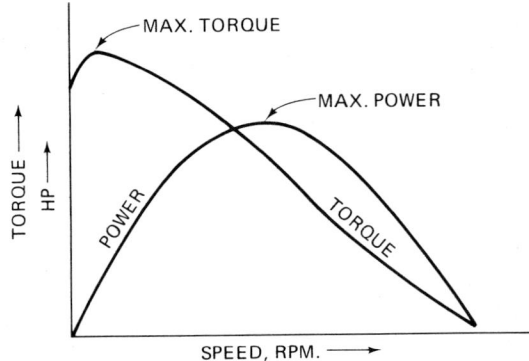

FIG. 22-2. Air motor horsepower and torque.

output. In practice, however, this is not so. Motor speed does affect the torque because of pressure losses in the air. Figure 22-2 shows typical torque and horsepower curves for air motors.

The hydrostatic transmission is a drive that includes a positive displacement pump and a motor. By using different combinations of variable displacement and fixed displacement pumps and motors, the various torque and horsepower curves shown in Fig. 22-3 can be obtained.

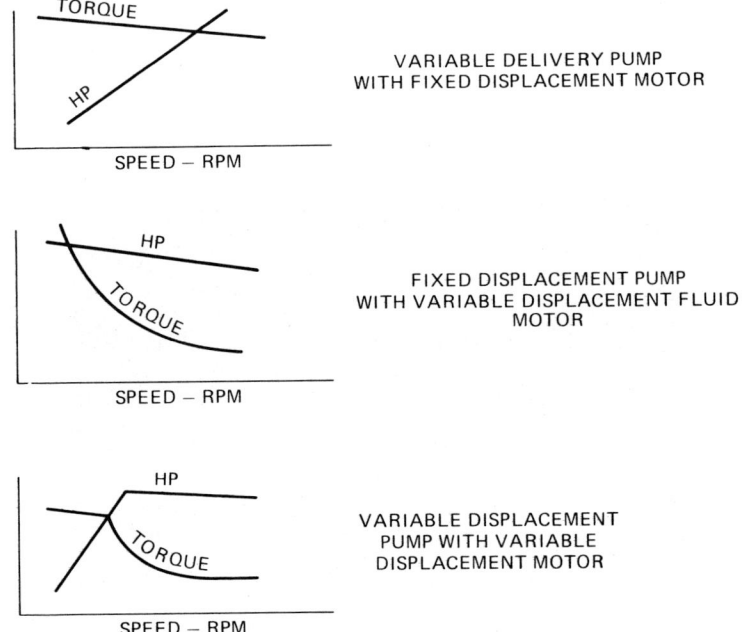

FIG. 22-3. Hydrostatic transmissions—torque and horsepower.

22-3 HYDRAULIC MOTORS

Figure 22-4 shows a high-torque, low-speed hydraulic motor manufactured by Sperry Vickers, a division of Sperry Rand Corporation. The theoretical torque output of this motor is 5000 lb ft at 2000 psi. Speed is 200 rpm maximum. The theoretical horsepower output of this motor is approximately 190 hp.

FIG. 22-4. Vickers MHT-250 hydraulic motor. (*Courtesy* Vickers division, Sperry Rand Corp.)

Figure 22-5 illustrates a smaller type of hydraulic motor. The design of these motors is termed modular, meaning that various motor outputs are available from one standard "package" size. The size of the motor is 4 inches square. The internal design of the motor is based on the use of a *gerotor*, a modified type of internal gear used for motors and pumps. The displacement of this type of motor may be obtained from 4 to 223 in.³ per revolution, with torque ranges up to 3000 lb in.

The similarity of design between the pump and motor has been men-

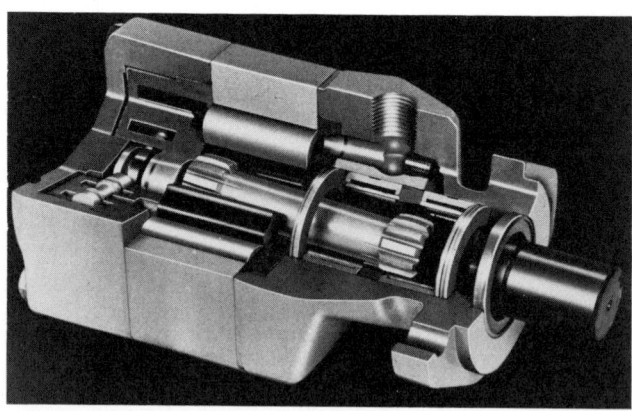

FIG. 22-5. High torque, low speed hydraulic motor. (*Courtesy* Vickers division, Sperry Rand Corp.)

Sect. 22-3 / HYDRAULIC MOTORS

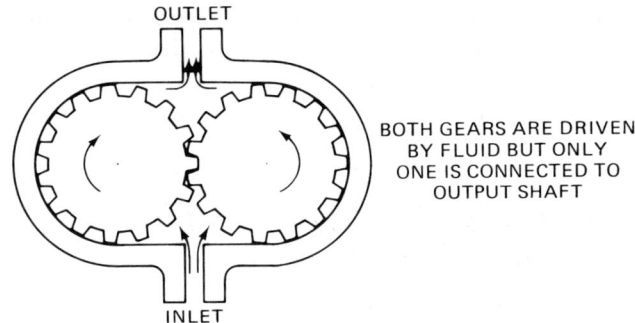

FIG. 22-6.   Gear motor.

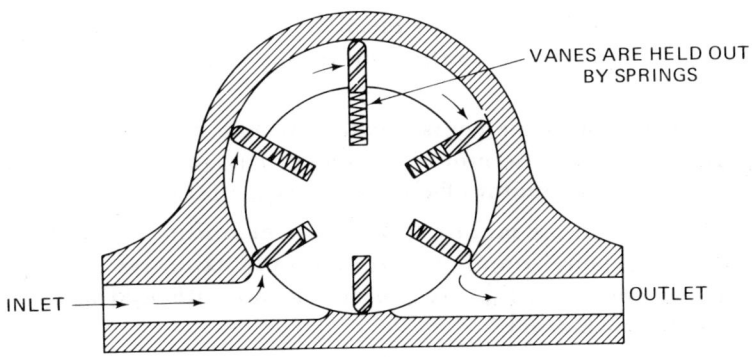

FIG. 22-7.   Vane type motor.

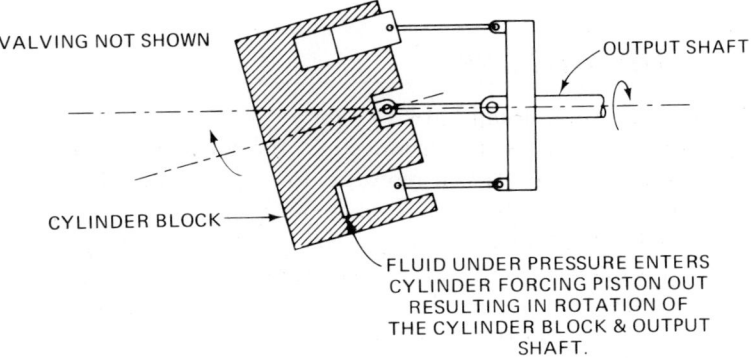

FIG. 22-8.   Axial piston motor.

tioned. Despite this similarity, there are differences which make a study of the general types of hydraulic motor designs desirable. As in pumps, three general types of positive-displacement motors are the gear, vane, and piston. Each of these types is shown and explained in Figs. 22-6, 22-7, and 22-8.

22-4 COMPARISON OF HYDRAULIC AND ELECTRIC MOTORS

Some comparisons of the hydraulic motor with the commonly used three-phase alternating current electric motor are interesting. A few are listed below.

1 / Power output of hydraulic motors for a given size is greater. Weight per horsepower for electric motor may range from 12 to 30 lb. Weight for hydraulic motors may range from $\frac{1}{2}$ to 5 lb per hp.
2 / Electric motor efficiency is greater. It may range from 90 to 95 percent. Hydraulic motor efficiency may range from 70 to 90 percent, and air motor efficiency is less than this.
3 / Hydraulic and air motors have more flexibility in operating in unfavorable environments. Operation in explosive atmospheres and submerged in liquids is easier to do with hydraulic rather than electric motors.
4 / Alternating-current motors have better speed regulation than either hydraulic or air motors, since the speed regulation of the electric motor is a function of the line frequency.
5 / The starting torque of the electric motor is higher.

22-5 AIR MOTORS

Gear, vane, and piston designs are used for air motors as well as hydraulic. In most cases, air motors are produced in the lower horsepower ranges. The power range available in air motors is from $\frac{1}{8}$ to 25 hp. Loaded speeds may range from 40 to 6000 rpm.

A special type of air motor is the *turbine*. An example of an application of this was mentioned in Chapter Sixteen (see Fig. 16-10). This is not a positive-displacement device, but it is useful where high speeds are desired. Figure 22-9 shows the design of a turbine motor.

Characteristics of the various types of air motors are described in Table 22-1.

Sect. 22-6 / HYDROSTATIC TRANSMISSIONS

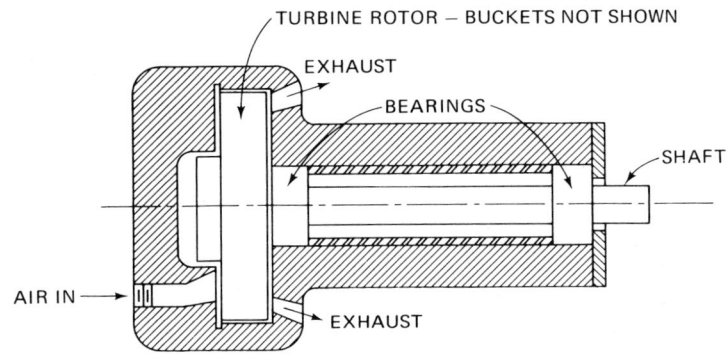

FIG. 22-9. Impulse turbine air motor.

Table 22-1   Some Characteristics of Air Motor Types

Type	Operating Pressure (psi)	Hp Range	Speed Range, (rpm)	Air Consumption (scfm)
Vane	80–90	0.3–10	800–6000	11–236
Axial piston	90	0.6–2.7	1300–2360	21–83
Radial piston	90–100	0.17–16	200–1500	6–400
Turbine	100	0.25–2	22000–50000	8–20

22-6   HYDROSTATIC TRANSMISSIONS

Performance curves for various hydrostatic transmissions are shown in Fig. 22-3. The *hydrostatic transmission,* also called a *hydrostatic drive,* is a combination unit consisting of pump and motor.

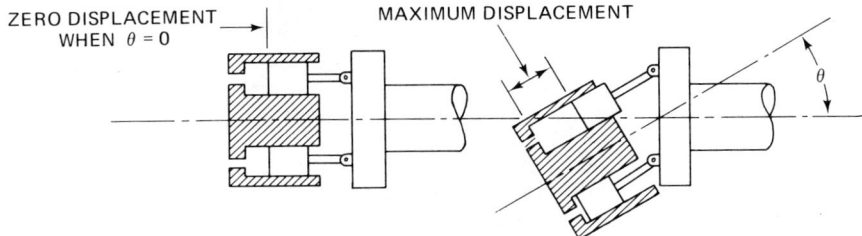

FIG. 22-10. Variable displacement principle for axial piston motors and pumps.

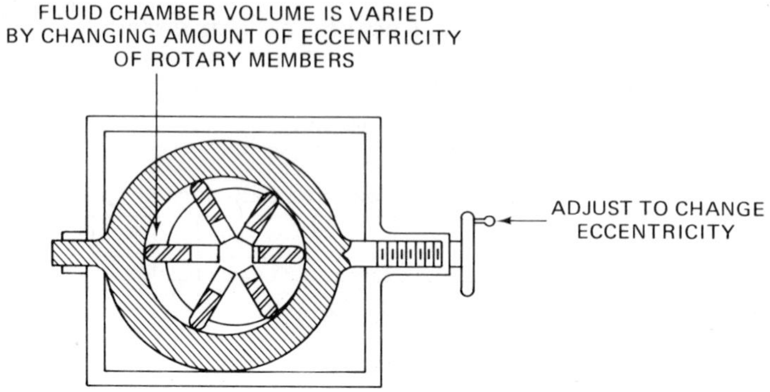

FLUID CHAMBER VOLUME IS VARIED
BY CHANGING AMOUNT OF ECCENTRICITY
OF ROTARY MEMBERS

ADJUST TO CHANGE
ECCENTRICITY

FIG. 22-11. Variable displacement principle for vane-type motors and pumps.

Figure 22-3 makes reference to variable-displacement pumps and motors. The variable-displacement pump or motor has a built-in adjusting mechanism and control for varying the volumetric displacement of the pump or motor. In the axial-piston variable-displacement pump this is accomplished by varying the angle between the drive shaft and the cylinder block. The principle is illustrated by Fig. 22-10. The principle used in

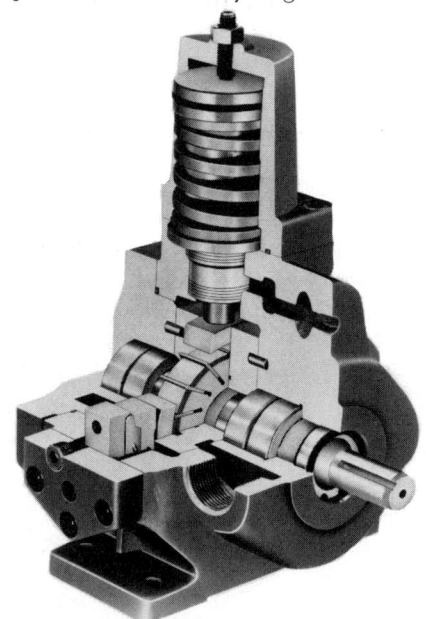

FIG. 22-12. Miller Fluid Power variable displacement vane pump. (*Courtesy* Miller Fluid Power, a Flick Reedy subsidiary Corp.)

varying the displacement of the vane-type pump or motor is shown in Fig. 22-11. The eccentric volume of the pump is varied in this case. Figure 22-12 shows a cutaway section of a Miller Fluid Power variable-displacement vane pump.

QUESTIONS AND PROBLEMS

22-1.—The displacement of an axial piston pump is 27 in.^{3} The actual output at 100 rpm is 2600 in.^{3}/min. What is the volumetric efficiency?

22-2.—Of the hydraulic and electric motor, which type has better speed regulation? What is the reason for the better speed regulation?

22-3.—What type of hydrostatic drive should be selected if it is required to have a constant torque over a range of speeds?

22-4.—Select a type of air motor to operate at 5 hp and 5000 rpm.

22-5.—What has greater power output per pound, the electric motor or the hydraulic motor?

22-6.—A hydraulic motor has a displacement of 54 in.^{3} When operating at 2000 rpm, the pressure drop across the motor is 40 psi. What is the output torque?

22-7.—In Problem 22-6, determine the horsepower of the motor.

22-8.—The hydrostatic transmission is a combination unit consisting of _____. (*Select the correct answer.*)

    a—Two pumps in series.

    b—A motor with torque regulation.

    c—A pump and a motor.

    d—A motor and a gear reducer.

22-9.—Which type of motor, hydraulic or electric, would be preferred for operation in an explosive atmosphere? Why?

22-10.—The characteristics of an air motor generally show the highest torque at the _____ speed ranges. (*Fill in the correct answer.*)

chapter twenty-three

ACCESSORY COMPONENTS

23-1 INTRODUCTION

In addition to the major components of the fluid power system there are other components which perform important functions. Some of these accessories have been mentioned briefly in preceding chapters but have not been explained in detail. This chapter goes into more detail on accessories, the need for them, and design details of some of them.

23-2 ACCUMULATOR

The *accumulator* is used in hydraulic systems as a stored source of pressurized fluid which is available for short-time use when a pump is off part of the time. A system such as this might work intermittently and require part-time pump power. The accumulator then provides a source of static pressure for a short time period until the pump comes back on the line.

A secondary use of the accumulator is to dampen line pressure shocks or surges in a hydraulic system. The familiar water hammer which occurs in a poorly designed home plumbing system is an example of this. This shock is caused by the too quick closing or operating of a valve or some other component in the system.

The bladder-type accumulator contains a rubber bladder in the larger shell of the accumulator. The bladder is charged with a gas such as nitrogen to a predetermined pressure. Air is not used because of the deleterious effects of oxygen on rubber over long periods of time. At the opposite end of the accumulator, fluid under pressure enters the shell. The gas pressure in the bladder increases and bladder volume decreases until the fluid pressure is equalized. At this point the system is in equilibrium.

As the fluid leaves the accumulator the pressure drops in the shell. The bladder then expands and maintains the gas pressure on the fluid. Functioning of the bladder-type accumulator is shown in Fig. 23-1.

The piston accumulator shown in Fig. 23-2 operates on the same principle as the bladder type. Separation between the gas and the incompressible fluid is accomplished by the piston instead of a bladder.

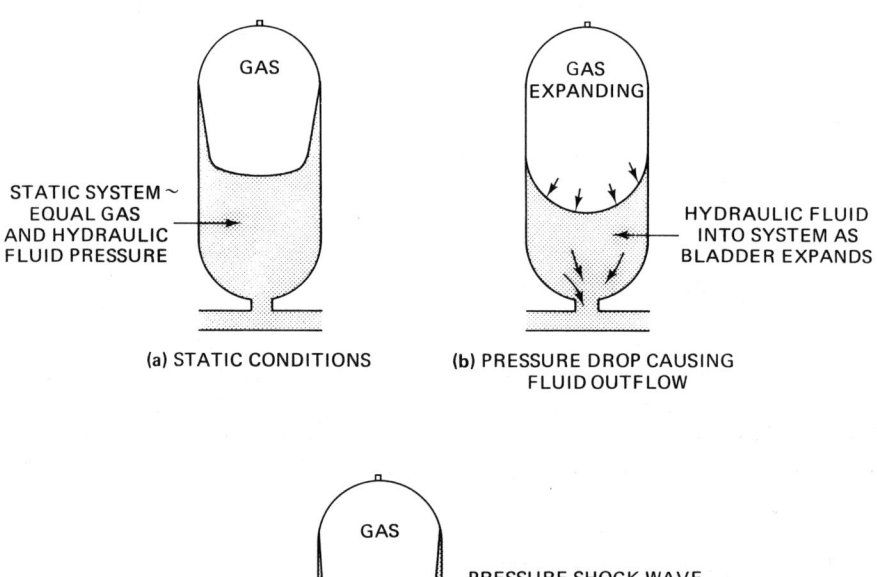

FIG. 23-1. Functioning of the gas bladder accumulator.

Sect. 23-3 / STRAINERS AND FILTERS

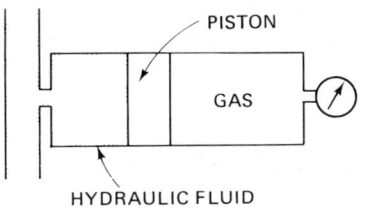

FIG. 23-2. Piston accumulator.

23-3 STRAINERS AND FILTERS

A JIC (Joint Industry Conference) definition of a *strainer* is a "device for the removal of solids from a fluid wherein the resistance to motion of such solids is in a straight line." The JIC definition of a *filter* is "a device for the removal of solids from a fluid wherein the resistance to motion of such solids is in a tortuous path."

An additional distinction between the two terms is the general design criterion that strainers are, as a rule, designed to remove coarser particles than filters. Frequently used together in a system, the strainer is placed in line first and the filter second.

Typical strainer construction is of mesh wire screens around a framework. In a hydraulic system the strainer usually is located in the pump intake in the reservoir fluid. A typical oil intake strainer installation is shown in Fig. 23-3.

Filter designs utilize numerous combinations of materials to achieve their purpose. These materials, called *media*, range from yarn, felt, and

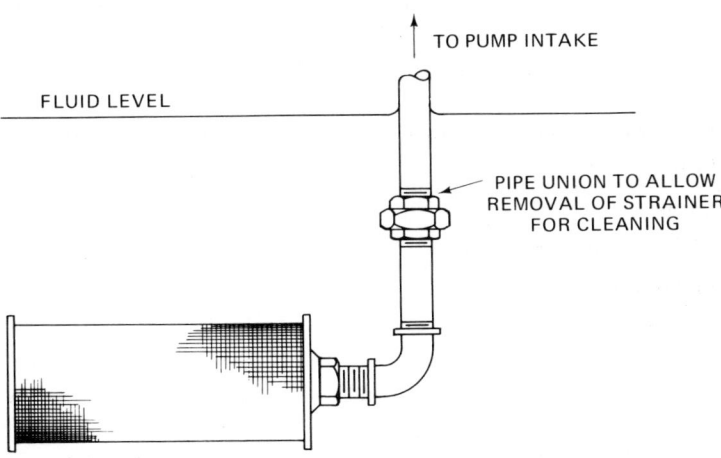

FIG. 23-3. Strainer installation.

cloth to sintered metals produced by powder metallurgy processes. While strainers find applications generally only with liquids, filters are used for both liquids and gases.

23-4 SPECIFYING A FILTER

Flow capacity, pressure drop across the filter, and particle size removed are three important considerations in specifying a filter. The manufacturer will normally rate a certain filter size for flow rate in gallons per minute of a fluid at a certain viscosity and with a certain pressure drop across the filter.

Particle removal rating may be based on a nominal removal method. This method specifies that 98 percent of all particles of a certain size will be removed by the filter. The size of the particle is usually specified in micrometers. One micrometer is equal to one millionth of a meter, or 0.000039 in. The micrometer is the correct term for particle-size measurement in the SI system. Prior to that, the micron was used to indicate one millionth of a meter. Because of its long-established usage, micron may still be seen in specifications and literature on filters. An idea of the sizes of the different particles can be obtained by examining Fig. 23-4. Figure 23-4 uses the term "micron" instead of micrometer, but the values are the same in both units.

The amount of contamination in a system should be considered when selecting the filter. All filter elements have a dirt capacity which, when reached, results in a higher than normal pressure drop or even blockage of the flow. To prevent complete blockage, bypass relief valving is provided in the filter. If it is desired to filter a dirty system to low levels of contamination, the filtration may have to be done in several stages. Using multistage filtration increases filter life to a satisfactory level.

Some of the media that are used for filter materials are:

1 / **Powder (sintered) metals such as bronze and stainless steel.**
2 / **Felt.**
3 / **Yarn.**
4 / **Wire mesh and cloth.**
5 / **Fritted glass.**
6 / **Paper**

Figure 23-5 shows a typical construction of a filter with a replaceable cylindrical cartridge.

Filters designed for air frequently are made to remove water as well as solids. This was discussed in Chapter Seventeen and a filter, regulator,

Sect. 23-4 / SPECIFYING A FILTER

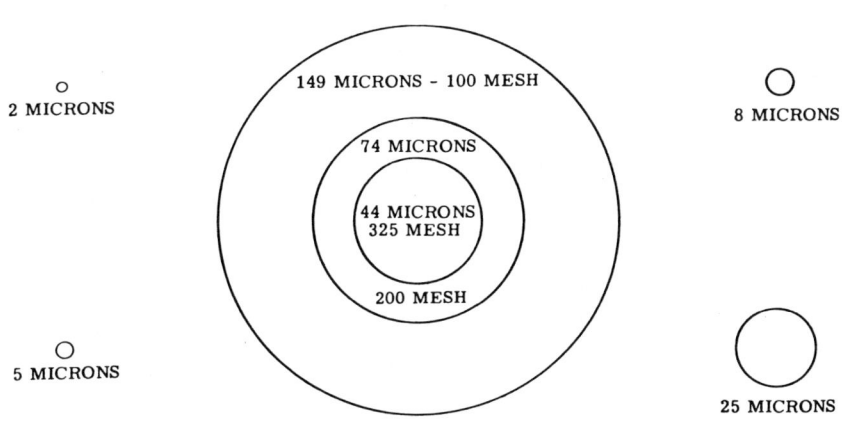

RELATIVE SIZES

LOWER LIMIT OF VISIBILITY (NAKED EYE)	40 MICRONS
WHITE BLOOD CELLS	25 MICRONS
RED BLOOD CELLS	8 MICRONS
BACTERIA (COCCI)	2 MICRONS

LINEAR EQUIVALENTS

1 INCH	25.4 MILLIMETERS		25,400 MICRONS
1 MILLIMETER	.0394 INCHES		1,000 MICRONS
1 MICRON	25,400 OF AN INCH		.001 MILLIMETERS
1 MICRON	3.94 x 10^{-5}		.000039 INCHES

SCREEN SIZES

MESHES PER LINEAR INCH	U.S. SIEVE NO.	OPENING IN INCHES	OPENING IN MICRONS
52.36	50	.0117	297
72.45	70	.0083	210
101.01	100	.0059	149
142.86	140	.0041	105
200.00	200	.0029	74
270.26	270	.0021	53
323.00	325	.0017	44
		.00039	10
		.000019	.5

FIG. 23-4. Relative size of micronic particles. Particle size is in microns instead of micrometers. (*Reproduced with permission, Sperry-Vickers, from Industrial Hydraulics Manual.*)

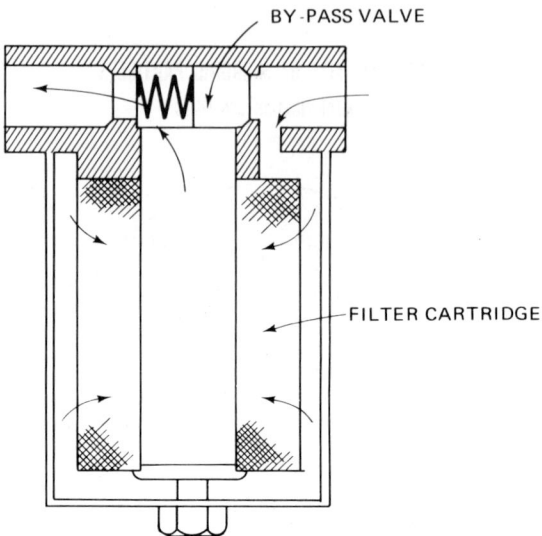

FIG. 23-5. Cartridge type filter.

and oiler combination unit were shown in Fig. 17-8. In Fig. 23-6 a cutaway view of an air filter on the left and an oiler on the right is shown. The filter element of this unit is sintered bronze and is located above the upside-down cup-shaped element in the bowl. The cup-shaped element serves as a slinger in separating water droplets from the air.

23-5 RESERVOIRS

Reservoirs for hydraulic systems can be selected as individual units or can be supplied as part of a complete power package. Such a power package is

FIG. 23-6. Air filter and oiler.

FIG. 23-7. Hydraulic power unit consisting of motor, pump and reservoir. (*Courtesy* Miller Fluid Power, a Flick-Reedy subsidiary.)

shown in Fig. 23-7. In this unit is contained the electric motor, pump, and reservoir. Sizing of the reservoir for the pump is already done when a unit such as this is specified.

Many variations of reservoir design may satisfy the demands of a hydraulic system. In general, the unit that operates satisfactorily and costs less is the best one for the system. The considerations below are among the most important in selecting a reservoir.

1 / The capacity of the reservoir should be 2 to 3 times the pump capacity as expressed in gpm.

2 / A vent to the atmosphere and tank drain should be provided.

3 / The vent should be equipped with an air breather filter.

4 / The tank should be baffled to provide for settling of contaminants and to prevent local turbulence which would interfere with pump suction.

5 / An oil-level gage should be provided.

6 / The capacity of the tank should be adequate to provide necessary cooling of the oil. An equation that may be used as a guide in determining the size for cooling is

$$0.001(\Delta T)(A) = HP \qquad (23\text{-}1)$$

where ΔT = temperature difference, °F, ambient vs. operating temperature

A = ft² of reservoir surface

HP = cooling capacity represented as horsepower; it can be converted to Btu/h by multiplying by 2544

7 / For operation in cold temperatures, immersion heaters, thermostatically controlled, may have to be provided.

Some of the considerations mentioned above are evident in the sketch of a typical reservoir in Fig. 23-8.

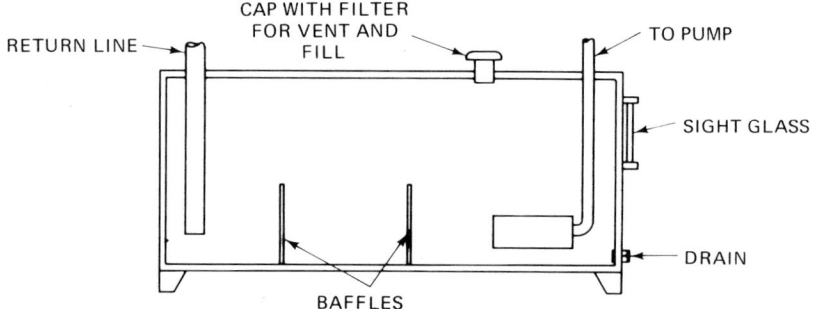

FIG. 23-8.   Reservoir.

QUESTIONS AND PROBLEMS

23-1.—Name two types of accumulators.

23-2.—Give five considerations in the design of a reservoir.

23-3.—A 10-micrometer filter should be able to remove approximately 98 percent of all particles down to _____ inches in size.

23-4.—Calculate the amount of surface area required for a reservoir to obtain 300 Btu/h cooling capacity if the ambient temperature is 70° F and the maximum desired oil temperature is 110° F.

23-5.—Define the difference between a strainer and a filter.

23-6.—Name three different materials used for filter elements.

chapter twenty-four

GRAPHIC SYMBOLS FOR FLUID POWER DIAGRAMS

24-1 INTRODUCTION

Circuit design in fluid power requires the use of the draftsman's language. Just as the electrical circuit requires special symbols, the fluid power circuit has its own symbols to specify the correct components.

In the past various organizations used different graphical standards. There now is an American National Standard, USAS Y32.10-1967, which contains standardized fluid power graphic symbols. The standardizing organization is the American National Standards Institute, formerly known as the American Standards Association. USAS Y32.10 is published by the American Society of Mechanical Engineers, with the collaboration of the National Fluid Power Association. All graphic symbols shown in this unit are extracted from USA Standard Graphic Symbols for Fluid Power Diagrams (USAS Y32.10-1967) with the permission of the publisher, the American Society of Mechanical Engineers, United Engineering Center, 345 East 47th Street, New York, New York 10017.

234　　　　GRAPHIC SYMBOLS FOR FLUID POWER DIAGRAMS / Chap. 24

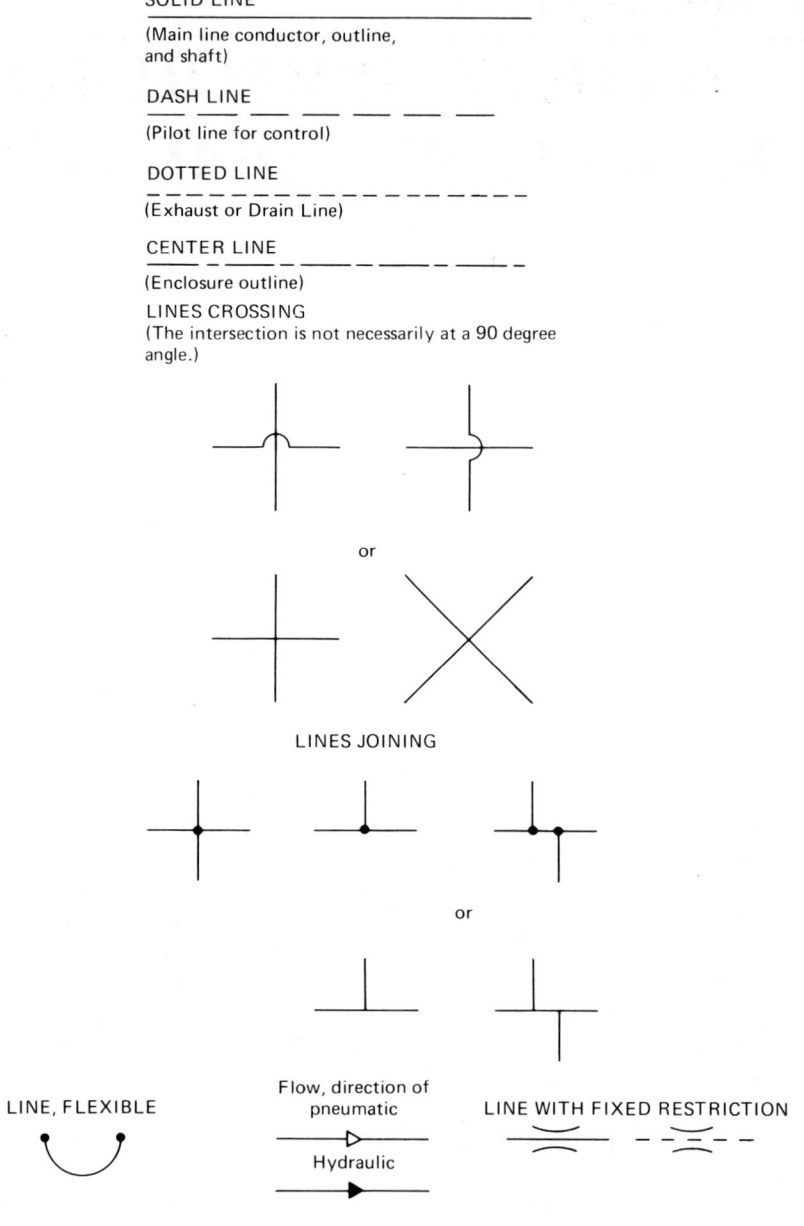

FIG. 24-1. Line usage.

Sect. 24-1 / INTRODUCTION

HYDRAULIC PUMP
FIXED DISPLACEMENT

Unidirectional

Bidirectional

HYDRAULIC MOTOR
FIXED DISPLACEMENT

Bidirectional

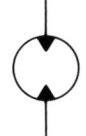

An arrow through a symbol at approximately 45 degrees indicates that the component can be adjusted or varied.

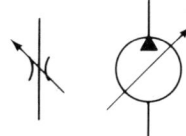

Rotating shafts are symbolized by an arrow which indicates direction of rotation (assume arrow on near side of shaft).

 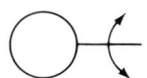

PUMP, PNEUMATIC

Compressor, fixed displacement

MOTOR, PNEUMATIC

Unidirectional

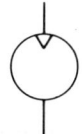

Vacuum pump, fixed displacement

Bidirectional

ELECTRIC MOTOR

FIG. 24-2. Pumps, motors, and compressors.

24-2 USE OF LINES

Figure 24-1 shows the basic line symbols and meanings. The distinction between the dashed line and the dotted line should be noted. In addition, note that the center line does not represent a hydraulic or pneumatic line carrying fluid. It represents a space or volume outline without a dimensional scale. Components located within such an outline would be assumed to be mounted reasonably close together and to function as a unit.

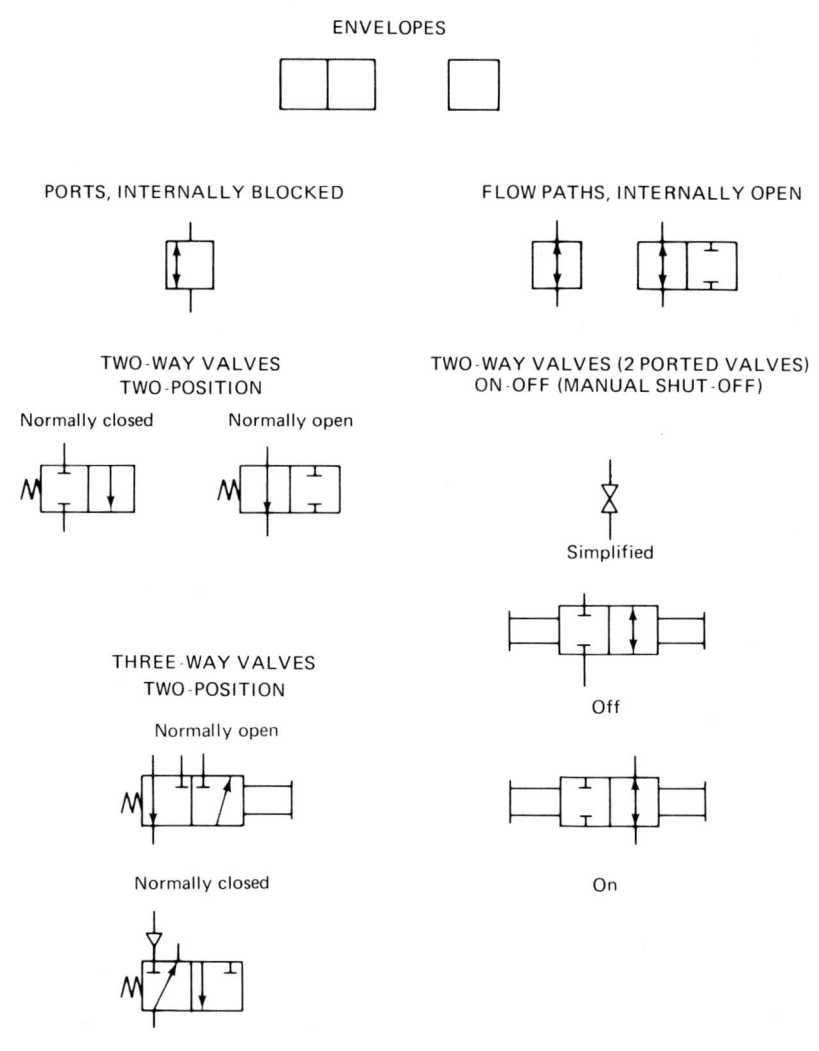

FIG. 24-3. Directional control valves—two-way and three-way.

24-3 SYMBOLS FOR COMPONENTS

The basic symbol for rotary power devices is the circle. It is used with various added features to indicate pumps, motors, and compressors. The variations of the symbols are shown in Fig. 24-2. Note that the only difference between the pump and motor symbol is the direction the arrowhead points with respect to the circumference of the circle.

The basic symbol for directional control valves is a square envelope which is increased in number as the valve complexity and functions increase. Figures 24-3 and 24-4 show the most common two-way, three-way,

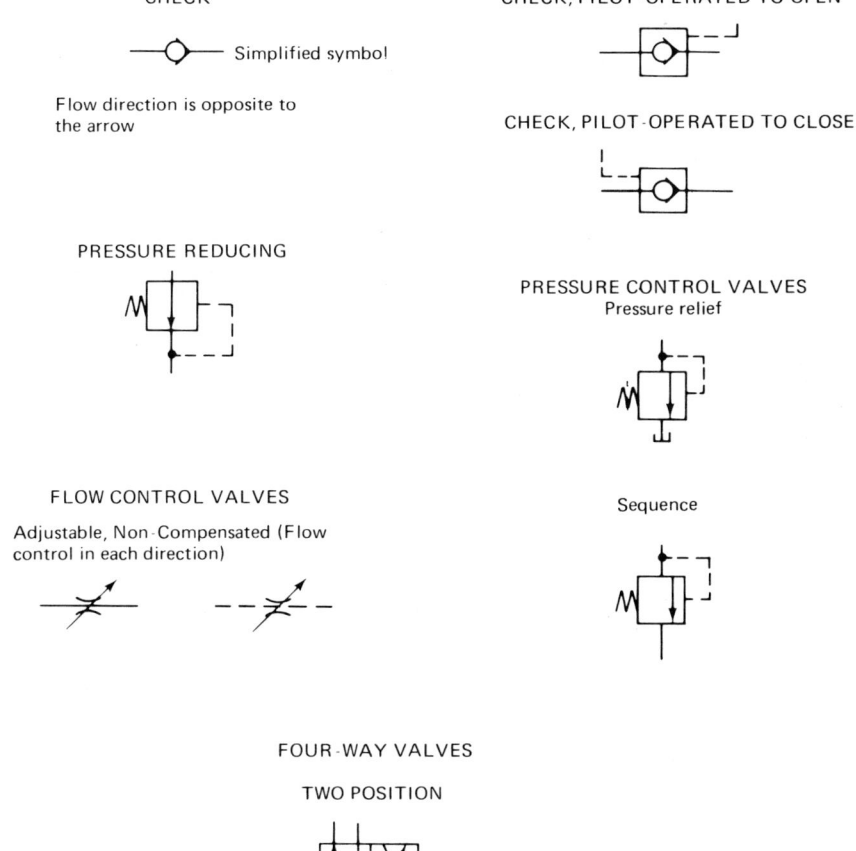

FIG. 24-4. Check valves, pressure relief, flow control, four-way directional and other valves.

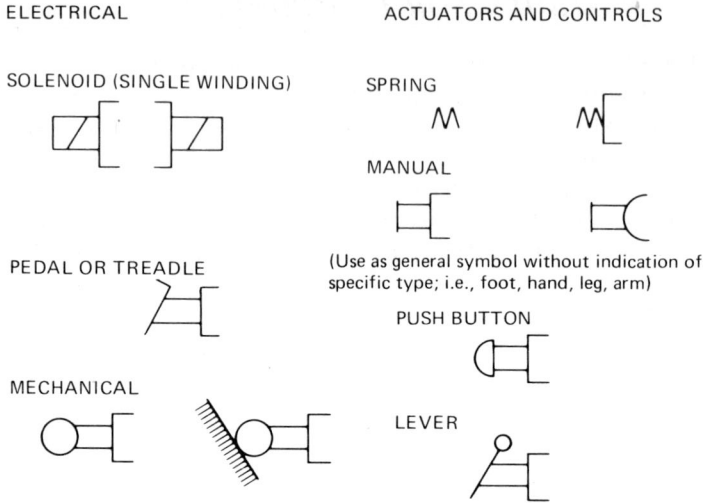

FIG. 24-5. Actuator symbols.

and four-way valve symbols. The symbols that are added at the ends of the envelope indicate how the valve is actuated. Actuation may be by lever, foot pedal, solenoid, or other means. The symbols representing these methods are shown in Fig. 24-5.

In Fig. 24-6 are shown the symbols used for reservoirs. Note that the simplified symbol may be used as the ground of an electrical circuit. It

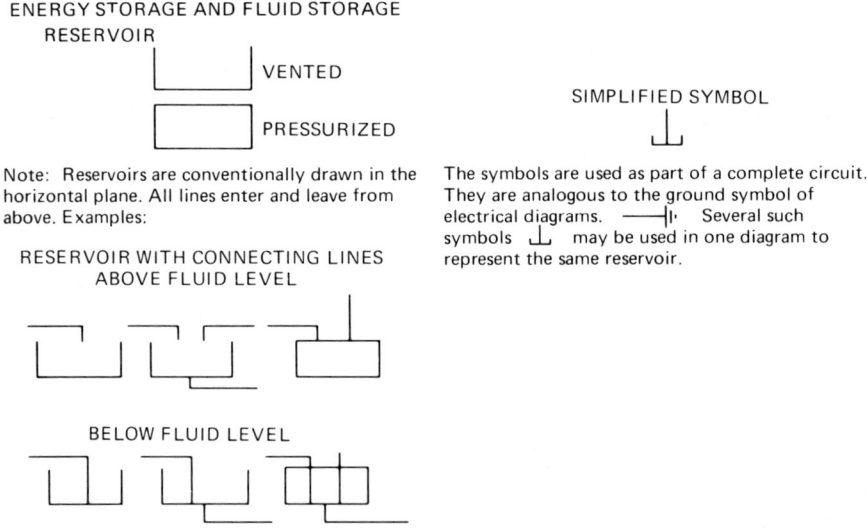

FIG. 24-6. Reservoir symbols.

Sect. 24-3 / SYMBOLS FOR COMPONENTS

ACCUMULATOR

FIG. 24-7. Accumulator.

would be used several times like this in one circuit, at any time a line returns to the reservoir.

Figure 24-7 shows the basic symbol for the accumulator. Variations of the symbol are provided for different types of accumulators.

The square standing on one corner is the symbol for the fluid conditioning devices. The major ones are shown in Fig. 24-8.

FLUID CONDITIONERS

FILTER - STRAINER

FILTER SEPARATOR
WITH AUTOMATIC DRAIN

SEPARATOR
WITH MANUAL DRAIN

DESICCATOR (CHEMICAL DRYER)

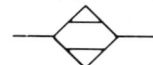

WITH AUTOMATIC DRAIN

LUBRICATOR
LESS DRAIN

FILTER - SEPARATOR
WITH MANUAL DRAIN

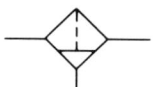

WITH MANUAL DRAIN

FIG. 24-8. Fluid conditioners.

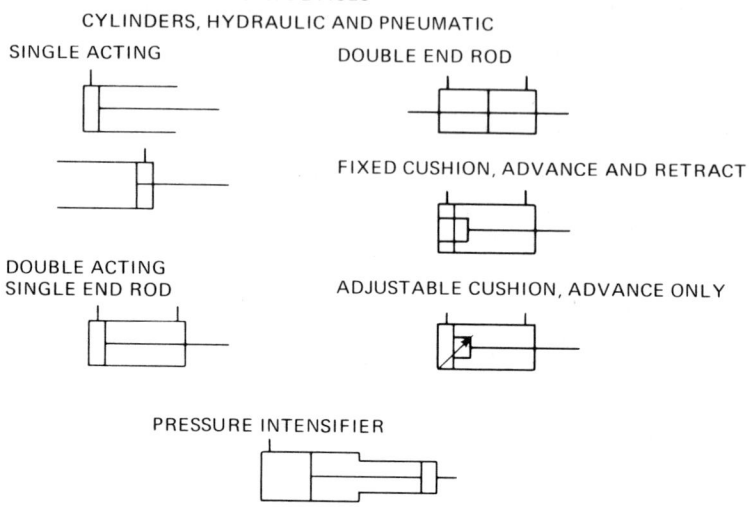

FIG. 24-9. Linear devices.

The linear devices, primarily cylinders, are represented by the symbols of Fig. 24-9.

24-4 OTHER SYMBOLS

The scope of this text prevents giving the complete set of graphic symbols. The ones shown are the major symbols and are adequate for simple circuits. For the complete set, the reader is referred to USAS Y32.10.

QUESTIONS AND PROBLEMS

24.1.—Draw symbols for the following:

  a—A solenoid-operated, two-position, three-way valve, normally closed.

  b—A lever-operated, two-way valve, spring return, normally closed.

  c—A fixed-displacement hydraulic motor.

  d—A reservoir with strainer.

  e—A variable line restriction.

QUESTIONS AND PROBLEMS

24-2.—Identify the following symbols.

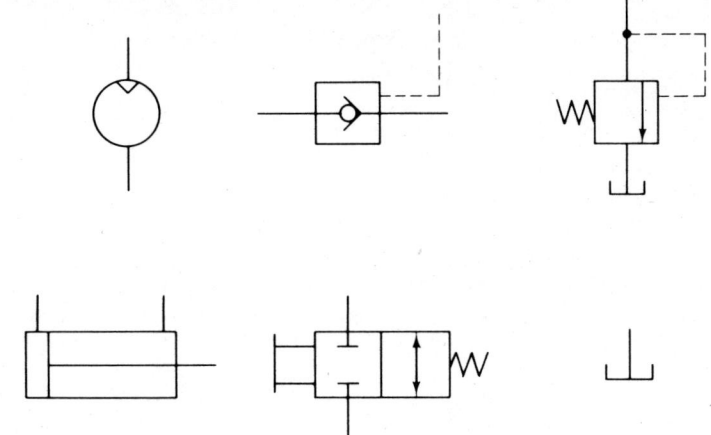

chapter twenty-five

HYDRAULIC CIRCUITS

25-1 INTRODUCTION

The hydraulic circuit ordinarily must consist of a pump and suitable reservoir as a starting point. Contained within the reservoir is a strainer and on the output side of the pump are suitable pressure relief and pressure control valves. Following these the actuators (cylinders, motors, etc.) are arranged so as to perform the desired functions. Proper design can allow for sequencing of the actuators, speed control, and timing of the functions. Although not within the scope of this text, electrical timing and sequencing devices offer a convenient way to operate direct solenoid-operated valves and solenoid-piloted valves in both hydraulic and pneumatic circuits.

Three simple hydraulic circuits are contained within this chapter. Each circuit is taken up separately for discussion.

25-2 SIMPLE PUMP CIRCUIT

The pump in this circuit (Fig. 25-1) pumps fluid from the reservoir through the strainer. On the pump output is a pressure relief valve which dumps excess fluid back into the reservoir when the pressure builds up to a preset point.

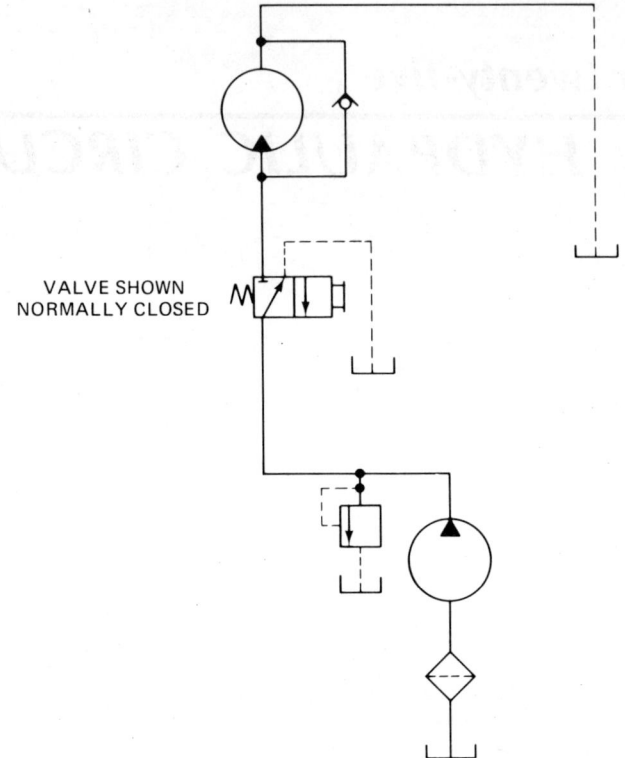

FIG. 25-1. Hydraulic motor operation by manual three-way valve.

A normally closed three-way valve is next in the circuit. Note that, in the closed portion as shown, the fluid is bypassed back into the reservoir. The valve is manually operated, as indicated by the right-hand portion on the valve symbol. When it is shifted open, fluid flows into the motor and rotates the motor. Fluid is returned to the reservoir from the motor.

The check valve in the bypass around the motor allows the motor to free-wheel to a stop when the inlet valve is closed. Fluid circulation around the bypass circuit is clockwise, and fluid is left in the pump inlet as the pump coasts to a stop. This feature prevents cavitation from occurring on startup.

25-3   *VARIABLE-OUTPUT PUMP CIRCUIT*

The pump, reservoir, and relief valve in this circuit (Fig. 25-2) are the same as Fig. 25-1, except that the pump has a variable output. This is indicated by the angled arrow through the pump symbol.

Sect. 25-4 / MOTOR-DRIVEN PUMP CIRCUIT

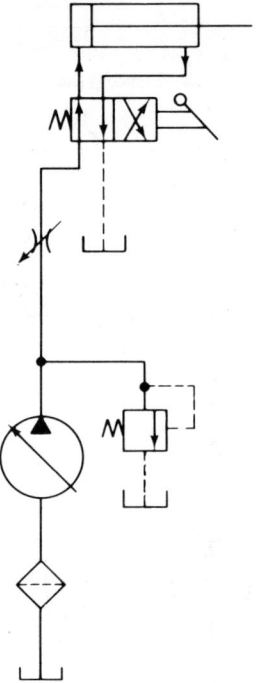

FIG. 25-2.   Four-way valve operating double acting cylinder.

The adjustable flow control valve next in the circuit controls the flow into the double-acting cylinder. In this position the flow control works when the cylinder is being extended and also when it is being retracted. The flow control valve reduces the speed at which the piston operates.

The four-way valve is manually operated by the lever. The drain from the cylinder returns to the reservoir through the valve. The pressure relief valve would have to be selected to function as a dumping valve as well as a maximum pressure relief valve.

25-4   MOTOR-DRIVEN PUMP CIRCUIT

An electric motor drive is shown (in Fig. 25-3) for the pump in this circuit. The flow control valve on the output side of the four-way valve controls the speed of the piston in one direction only. The speed control operates as the piston moves to the right. The check valve allows free flow as the fluid is exhausted from the left side of the cylinder.

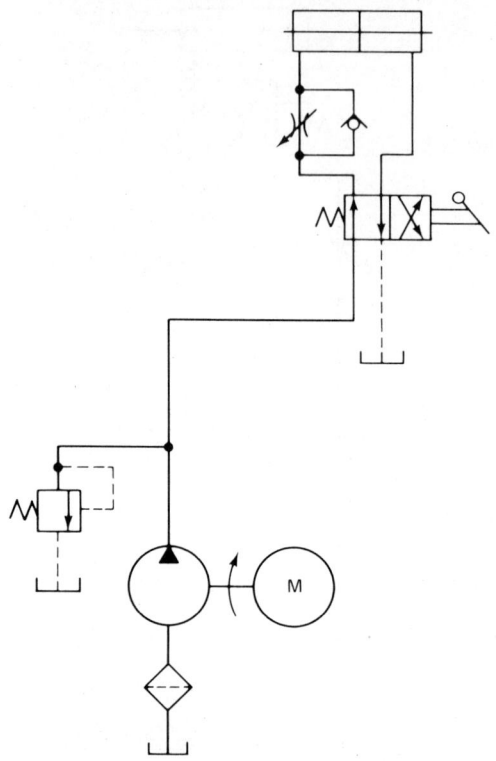

FIG. 25-3. Speed control of a hydraulic cylinder in one direction.

QUESTIONS AND PROBLEMS

25-1.—In Fig. 25-1, add a double-acting cylinder to the circuit. The cylinder is to be operated by a four-way valve operated by solenoid.

25-2.—In Fig. 25-2, redesign the circuit to provide speed control on the retract cycle only of the cylinder.

25-3.—Provide speed control in both directions for the cylinder in Fig. 25-3.

25-4.—In Fig. 25-4, add the appropriate components in the dashed circular areas A, B, C, D, and E, to accomplish the following functions:

QUESTIONS AND PROBLEMS

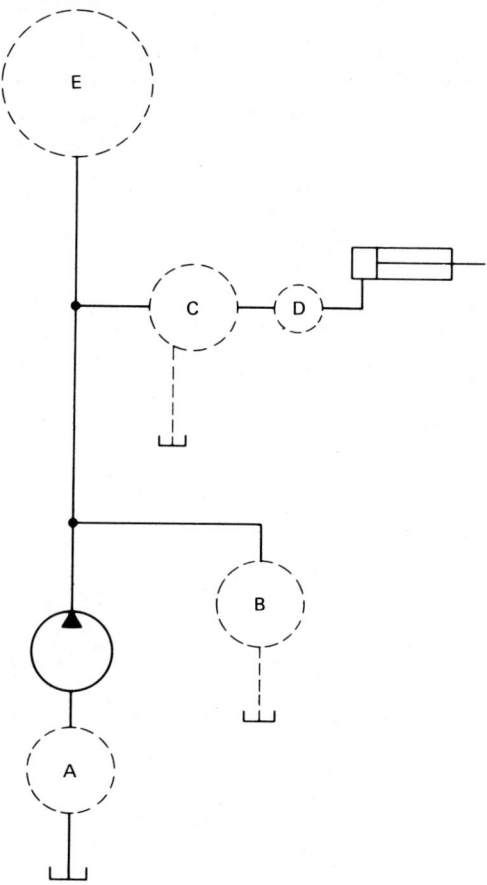

FIG. 25-4.

a—Fluid conditioning into the pump.
b—Pressure bypass back into the reservoir.
c—Directional control valve for cylinder.
d—Speed control for cylinder.
e—Double-acting cylinder with appropriate valve(s).

chapter twenty-six
PNEUMATIC CIRCUITS

26-1 INTRODUCTION

The design of pneumatic circuits usually starts with conditioning the air already available from a shop air supply. This is not true in all cases, as the air suspension system and air braking systems of vehicles have to be provided with air compressors.

Unlike the hydraulic system, the pneumatic system works directly from a storage tank or receiver. The compressor then has the job of keeping the receiver filled with air at the proper pressure. In most plants a range of pressure, approximately 100 to 120 psi, is maintained in the receiver. Automatic cut-in of the compressor is provided when the receiver pressure drops to the low point. When the pressure reaches a maximum the compressor cuts out.

Figures 26-1 through 26-4 show circuits and portions of circuits. Each one is discussed separately.

26-2 COMPLETE PNEUMATIC CIRCUIT

Figure 26-1 shows the compressor and storage tank in a complete circuit. Note that filtration is provided at the intake of the compressor. The air is

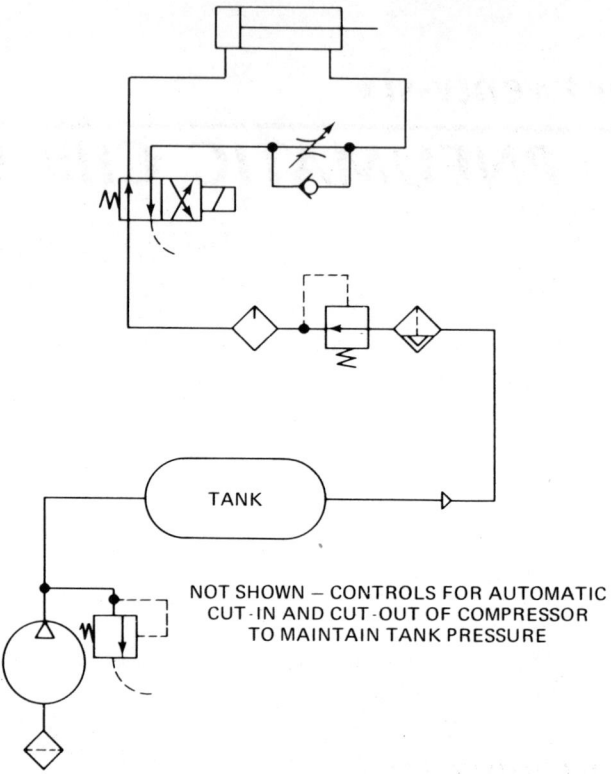

FIG. 26-1. Pneumatic circuit for cylinder speed control.

conditioned after the tank by a filter and separator and a lubricator. A pressure-reducing valve is provided between the filter and lubricator. This valve also is a regulator and stabilizes the pressure downstream from this point.

The double-acting cylinder operation is similar to that in hydraulics, except that exhaust is to atmosphere.

26-3 TWO-CYLINDER PARTIAL CIRCUIT

The two-cylinder partial circuit in Fig. 26-2 is designed so as to allow cylinder 1 to operate before cylinder 2 does. The flow on cylinder 2 slows it down enough so that cylinder 1 can complete its stroke. The circuit is used for clamping a part in a machine and then performing work on it. Cylinder 1 is the clamp cylinder and cylinder 2 does the work.

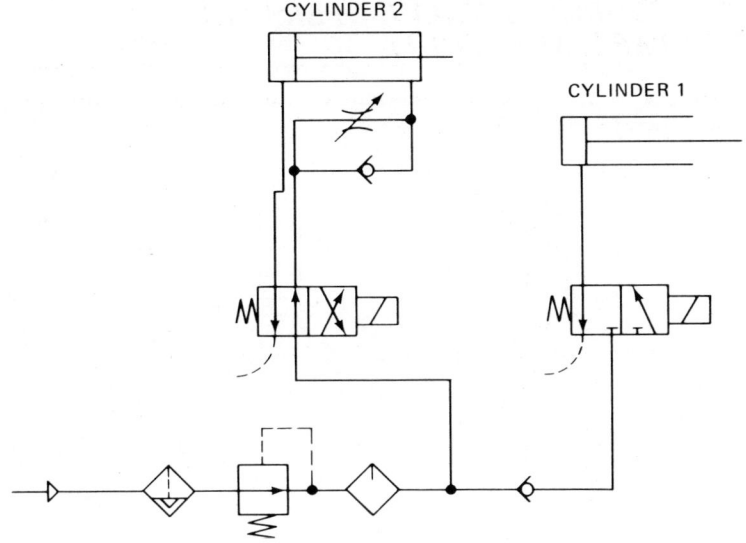

FIG. 26-2. Two cylinder circuit.

26-4 VALVE-OPERATED PARTIAL CIRCUIT

The circuit of Fig. 26-3 shows how two cylinders may be operated in parallel by one valve. The only way to make the cylinders operate at the same time is to connect the rods mechanically since the speed of the cylinders cannot be controlled accurately.

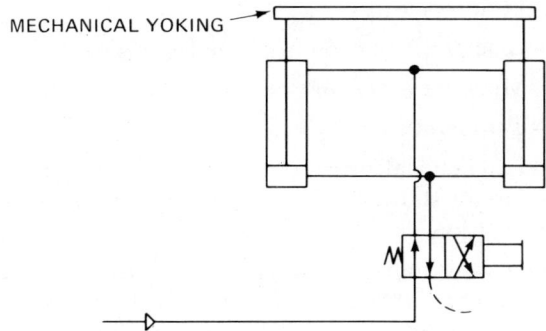

FIG. 26-3. Two cylinders in parallel.

26-5 FOUR-WAY VALVE-OPERATED PARTIAL CIRCUIT

Each cylinder is operated independently and requires its own four-way valve, as shown in Fig. 26-4.

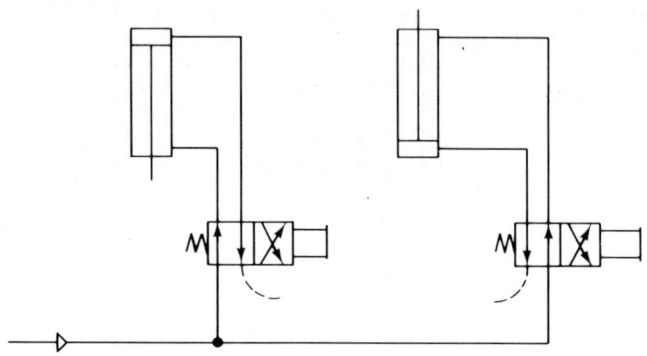

FIG. 26-4. Each cylinder with individual operation requires a four-way valve.

QUESTIONS AND PROBLEMS

26-1.—Describe the differences in conditioning the fluid for a hydraulic and a pneumatic circuit.

26-2.—In a large manufacturing plant _____ frequently becomes an important factor to consider when installing pneumatic equipment. (*Select the correct answer.*)

  a—Fluid viscosity.

  b—Line pressure drop from the compressor.

  c—Moisture in the air.

  d—Temperature.

26-3.—Add a chemical dryer (desiccator) in the proper location for the circuit shown in Fig. 26-3. Refer to Fig. 24-8 for further information.

26-4.—The circuit in Fig. 26-4 is supplied from a large air receiver tank kept at 200 psi. Sketch the portion of the circuit from the receiver to the cylinders, showing the other circuit components necessary. The cylinders are to operate on 100 psi pressure. Air in the receiver is assumed to contain moisture and dirt.

26-5.—Design and sketch a pneumatic circuit showing a compressor operating a single-action spring-return cylinder and a unidirectional motor. The motor and cylinder are to be individually controlled by solenoid-operated valves. Motor operating pressure is 55 to 65 psi and the cylinder operates at 90 to 100 psi. The compressor output is 140 to 150 psi. Show all necessary valving, fluid conditioners, and pressure regulators.

appendix one

SUPPLEMENTARY TABLES

Table I Properties of Some Common Liquids

Liquid	Temp. (°F)	Specific Gravity	Specific Weight (lb/ft^3)	Vapor Pressure (psia)
Alcohol, ethyl	68	0.789	49.2	0.85 at 68°F
Alcohol, methyl	70	0.796	49.7	1.965 at 70° F
Benzene	68	0.879	54.8	1.448 at 68° F
Carbon tetrachloride	68	1.595	99.5	14.7 at 170° F
Castor oil	68	0.960	59.9	—
Gasoline	60	0.68–0.74	42.2–46.2	4.42 at 70° F
Glycerine	68	1.262	78.7	—
Kerosene	60	0.82	51.2	0.492 at 70° F
Linseed oil	60	0.942	58.8	—
Mercury	39	13.6	848.1	0.000025 at 70° F
Mineral oil	68	0.88–0.94	54.9–58.7	—
Motor oil—SAE 10, 30, 50	60	0.91	56.8	—
Phosphate ester oil	68	1.1	68.6	0.004 to 0.008 at 248° F
Seawater	60	1.03	64.3	—
Silicone oil—F60	77	1.04	64.9	Less than 0.02 at 400° F
Water	50	1	62.4	0.339 at 68° F
Water-glycol solution	68	1.1	68.6	—
Water–oil emulsion	60	0.92–1.06	57.4–66.1	—

Table II   Equivalent Units in the English and SI Systems

Force Units

lb force	newton
1	4.4482
0.2248	1

Mass Units

gram	kilogram	lb (avoirdupois)
1	0.001	0.0022
1000	1	2.2046
453.6	0.4536	1

Length Units

millimeter	centimeter	meter	inch	foot
1	0.1	0.001	0.03937	0.00328
10	1	0.01	0.3937	0.0328
1000	100	1	39.37	3.281
25.4	2.54	0.0254	1	0.0833
304.8	30.48	0.3048	12	1

Volume Units

liter	U.S. gallon	cubic foot
1	0.2642	0.0353
3.785	1	0.1337
28.32	7.48	1

Pressure Units

pascal (Pa)	kilopascal (kPa)	lb/in.^{2}	bar*
1	0.001	0.000145	10^{-5}
1000	1	0.145	0.01
6894.8	6.8948	1	0.0689
10^5	100	14.5	1

Energy Units

Btu	joule (J)	ft lb
1	1055	778
0.0009478	1	0.7376
0.001285	1.3558	1

Power Units

horsepower	watt (W)	ft lb/s	Btu/s
1	746	550	0.7068
0.001341	1	0.7376	0.00095
0.00182	1.356	1	0.001285
1.415	1055	778	1

*The bar is a metric unit of pressure which was approved in 1969 for use for a limited time period. Because it is a larger unit than the pascal, many authorities in the fluid power industry use it in preference to the pascal and kilopascal, even though it is not an SI unit.

Table III Kinematic Viscosity of Various Liquids at Different Temperatures

A. (Multiply tabular value by 10^{-4} to obtain kinematic viscosity in $ft^2/sec.$)

TEMP. °F	10	30	50	70	90	110	150	200
Crude oil, $S_g = 0.855$	6	1.8	1.2	0.75	0.58	0.44	0.35	0.24
Kerosene, $S_g = 0.813$	–	0.48	0.31	0.23	0.18	0.16	0.13	0.1
Gasoline, $S_g = 0.716$	–	0.09	0.07	0.06	0.055	0.05	0.043	–
SAE 10 Lube oil	–	–	–	9.7	5	3	1.5	0.75
SAE 30 Lube oil	–	–	–	29	15	8.3	3.7	1.5
Water	–	–	0.14	0.11	0.08	0.07	0.05	0.03
Medium lube oil – $S_g = 0.89$	–	–	28	12.5	6.9	3.8	1.6	–
Medium fuel oil – $S_g = 0.85$	–	–	0.555	0.41	0.32	0.23	–	–
Heavy fuel oil – $S_g = 0.91$	–	–	31.2	15.7	8.4	4.8	–	–

B. Viscosities of other liquids — units and temperatures as indicated

Water-glycol solutions	20°F	40°F	100°F	130°F	
Ucon Hydrolube AC	–	23 cs	917 cs	9.0 cs	
Ucon Hydrolube 200-N	1535 SSU	806 SSU	201 SSU	124 SSU	
Ucon Hydrolube 300-N	2460 SSU	1280 SSU	306 SSU	174 SSU	
Water-oil emulsions	40°F	70°F	100°F	140°F	180°F
Shell Iris Fluid 902	2200 SSU	800 SSU	365 SSU	167 SSU	88 SSU
Phosphate esters	–40°F	100°F	210°F		
Skydrol 7000	7000 cs	15.7 cs	4 cs		

Table IV Gas Properties

Gas	Gas Constant R (ft lb/lb °R)	Molecular Weight (m)	Adiabatic Exponent n	Specific Weight at 1 Atm and 68° F (lb/ft³)
Air	53.3	29	1.4	0.0753
Ammonia	89.5	17	1.29	0.0442
Carbon dioxide	34.9	44	1.28	0.1142
Hydrogen	767	2	1.4	0.0052
Methane	96.3	16	1.32	0.0416
Nitrogen	55.1	28	1.4	0.0727
Oxygen	48.3	32	1.4	0.0830

Table V Common Equivalents

Volumes
  1 gal = 4 qt = 0.1337 ft³ = 231 in.³
  1 ft³ = 1728 in.³ = 7.48 gal = 29.92 qt

Energy
  1 Btu = 778 ft lb
  1 hp-h = 2544 Btu = 1.98 × 10⁶ ft lb
  1 kWh = 3412 Btu = 2.655 × 10⁶ ft lb

Power
  1 hp = 746 W = 550 $\frac{\text{ft lb}}{\text{s}}$ = 33,000 $\frac{\text{ft lb}}{\text{min}}$ = 42.4 $\frac{\text{Btu}}{\text{min}}$

Pressure
  1 atm = 14.7 psi = 29.92 in. Hg = 33.93 ft water
  1 psi = 0.06805 atm = 2.036 in. Hg = 2.309 ft water
  1 in. Hg = 0.491 psi = 0.0334 atm = 1.134 ft water
  1 ft water = 0.02947 atm = 0.433 psi = 0.8819 in. Hg

Table VI Properties of Steel Pipe*

Nominal pipe size, O.D., in.	Schedule number a	Schedule number b	Schedule number c	Wall thickness, in.	I.D., in.	Inside area, sq in.	Metal area, sq in.	Sq ft. outside surface, per ft	Sq ft. inside surface, per ft	Weight per ft. lb
1/8 0.405	... 40 80	..... Std XS	10S 40S 80S	0.049 0.068 0.095	0.307 0.269 0.215	0.0740 0.0568 0.0364	0.0548 0.0720 0.0925	0.106 0.106 0.106	0.0804 0.0705 0.0563	0.186 0.245 0.315
1/4 0.540	... 40 80	..... Std XS	10S 40S 80S	0.065 0.088 0.119	0.410 0.364 0.302	0.1320 0.1041 0.0716	0.0970 0.1250 0.1574	0.141 0.141 0.141	0.1073 0.0955 0.0794	0.330 0.425 0.535
3/8 0.675	... 40 80	..... Std XS	10S 40S 80S	0.065 0.091 0.126	0.545 0.493 0.423	0.2333 0.1910 0.1405	0.1246 0.1670 0.2173	0.177 0.177 0.177	0.1427 0.1295 0.1106	0.423 0.568 0.739
1/2 0.840	... 40 80 160 ...	..... Std XS ..... XXS	10S 40S 80S	0.083 0.109 0.147 0.187 0.294	0.674 0.622 0.546 0.466 0.252	0.357 0.304 0.2340 0.1706 0.0499	0.1974 0.2503 0.320 0.383 0.504	0.220 0.220 0.220 0.220 0.220	0.1765 0.1628 0.1433 0.1220 0.0660	0.671 0.851 1.088 1.304 1.714
3/4 1.050	... ... 40 80 160 ...	.... Std XS ..... XXS	5S 10S 40S 80S	0.065 0.083 0.113 0.154 0.218 0.308	0.920 0.884 0.824 0.742 0.614 0.434	0.665 0.614 0.533 0.432 0.2961 0.1479	0.2011 0.2521 0.333 0.435 0.570 0.718	0.275 0.275 0.275 0.275 0.275 0.275	0.2409 0.2314 0.2157 0.1943 0.1607 0.1137	0.684 0.857 1.131 1.474 1.937 2.441
1 1.315	... ... 40 80 160 ...	.... Std XS ..... XXS	5S 10S 40S 80S	0.065 0.109 0.133 0.179 0.250 0.358	1.185 1.097 1.049 0.957 0.815 0.599	1.103 0.945 0.864 0.719 0.522 0.2818	0.2553 0.413 0.494 0.639 0.836 1.076	0.344 0.344 0.344 0.344 0.344 0.344	0.310 0.2872 0.2746 0.2520 0.2134 0.1570	0.868 1.404 1.679 2.172 2.844 3.659
1¼ 1.660	... ... 40 80 160 ...	.... Std XS ..... XXS	5S 10S 40S 80S	0.065 0.109 0.140 0.191 0.250 0.382	1.530 1.442 1.380 1.278 1.160 0.896	1.839 1.633 1.496 1.283 1.057 0.631	0.326 0.531 0.669 0.881 1.107 1.534	0.434 0.434 0.434 0.434 0.434 0.434	0.401 0.378 0.361 0.335 0.304 0.2346	1.107 1.805 2.273 2.997 3.765 5.214
1½ 1.900	... ... 40 80 160 ...	.... Std XS ..... XXS	5S 10S 40S 80S	0.065 0.109 0.145 0.200 0.281 0.400	1.770 1.682 1.610 1.500 1.338 1.100	2.461 2.222 2.036 1.767 1.406 0.950	0.375 0.613 0.799 1.068 1.429 1.885	0.497 0.497 0.497 0.497 0.497 0.497	0.463 0.440 0.421 0.393 0.350 0.288	1.274 2.085 2.718 3.631 4.859 6.408
2 2.375	... ... 40 80 160 ...	.... Std XS ..... XXS	5S 10S 40S 80S	0.065 0.109 0.154 0.218 0.343 0.436	2.245 2.157 2.067 1.939 1.689 1.503	3.96 3.65 3.36 2.953 2.240 1.774	0.472 0.776 1.075 1.477 2.190 2.656	0.622 0.622 0.622 0.622 0.622 0.622	0.588 0.565 0.541 0.508 0.442 0.393	1.604 2.638 3.653 5.022 7.444 9.029

† a. ASA B36.10 steel pipe schedule numbers
  b. ASA B36.10 steel pipe nominal wall thickness designations
  c. ASA B36.19 stainless steel pipe schedule numbers

*Reproduced with permission of the ITT Grinnell Corporation.

Table VI Properties of Steel Pipe (Cont'd)

Nominal pipe size, O.D., in.	Schedule number† a	b	c	Wall thickness, in.	I.D., in.	Inside area, sq in.	Metal area, sq in.	Sq ft. outside surface, per ft	Sq ft. inside surface, per ft	Weight per ft. lb
2½ 2.875	...	.....	5S	0.083	2.709	5.76	0.728	0.753	0.709	2.475
	...	.....	10S	0.120	2.635	5.45	1.039	0.753	0.690	3.531
	40	Std	40S	0.203	2.469	4.79	1.704	0.753	0.646	5.793
	80	XS	80S	0.276	2.323	4.24	2.254	0.753	0.608	7.661
	160	.....	...	0.375	2.125	3.55	2.945	0.753	0.556	10.01
	...	XXS	...	0.552	1.771	2.464	4.03	0.753	0.464	13.70
3 3.500	...	.....	5S	0.083	3.334	8.73	0.891	0.916	0.873	3.03
	...	.....	10S	0.120	3.260	8.35	1.274	0.916	0.853	4.33
	40	Std	40S	0.216	3.068	7.39	2.228	0.916	0.803	7.58
	80	XS	80S	0.300	2.900	6.61	3.02	0.916	0.759	10.25
	160	.....	...	0.437	2.626	5.42	4.21	0.916	0.687	14.32
	...	XXS	...	0.600	2.300	4.15	5.47	0.916	0.602	18.58
3½ 4.000	...	.....	5S	0.083	3.834	11.55	1.021	1.047	1.004	3.47
	...	.....	10S	0.120	3.760	11.10	1.463	1.047	0.984	4.97
	40	Std	40S	0.226	3.548	9.89	2.680	1.047	0.929	9.11
	80	XS	80S	0.318	3.364	8.89	3.68	1.047	0.881	12.51
4 4.500	...	.....	5S	0.083	4.334	14.75	1.152	1.178	1.135	3.92
	...	.....	10S	0.120	4.260	14.25	1.651	1.178	1.115	5.61
	40	Std	40S	0.237	4.026	12.73	3.17	1.178	1.054	10.79
	80	XS	80S	0.337	3.826	11.50	4.41	1.178	1.002	14.98
	120	.....	...	0.437	3.626	10.33	5.58	1.178	0.949	18.96
	160	.....	...	0.531	3.438	9.28	6.62	1.178	0.900	22.51
	...	XXS	...	0.674	3.152	7.80	8.10	1.178	0.825	27.54
5 5.563	...	.....	5S	0.109	5.345	22.44	1.868	1.456	1.399	6.35
	...	.....	10S	0.134	5.295	22.02	2.285	1.456	1.386	7.77
	40	Std	40S	0.258	5.047	20.01	4.30	1.456	1.321	14.62
	80	XS	80S	0.375	4.813	18.19	6.11	1.456	1.260	20.78
	120	.....	...	0.500	4.563	16.35	7.95	1.456	1.195	27.04
	160	.....	...	0.625	4.313	14.61	9.70	1.456	1.129	32.96
	...	XXS	...	0.750	4.063	12.97	11.34	1.456	1.064	38.55
6 6.625	...	.....	5S	0.109	6.407	32.2	2.231	1.734	1.677	5.37
	...	.....	10S	0.134	6.357	31.7	2.733	1.734	1.664	9.29
	40	Std	40S	0.280	6.065	28.89	5.58	1.734	1.588	18.97
	80	XS	80S	0.432	5.761	26.07	8.40	1.734	1.508	28.57
	120	.....	...	0.562	5.501	23.77	10.70	1.734	1.440	36.39
	160	.....	...	0.718	5.189	21.15	13.33	1.734	1.358	45.30
	...	XXS	...	0.864	4.897	18.83	15.64	1.734	1.282	53.16
8 8.625	...	.....	5S	0.109	8.407	55.5	2.916	2.258	2.201	9.91
	...	.....	10S	0.148	8.329	54.5	3.94	2.258	2.180	13.40
	20	.....	...	0.250	8.125	51.8	6.58	2.258	2.127	22.36
	30	.....	...	0.277	8.071	51.2	7.26	2.258	2.113	24.70
	40	Std	40S	0.322	7.981	50.0	8.40	2.258	2.089	28.55
	60	.....	...	0.406	7.813	47.9	10.48	2.258	2.045	35.64
	80	XS	80S	0.500	7.625	45.7	12.76	2.258	1.996	43.39
	100	.....	...	0.593	7.439	43.5	14.96	2.258	1.948	50.87
	120	.....	...	0.718	7.189	40.6	17.84	2.258	1.882	60.63
	140	.....	...	0.812	7.001	38.5	19.93	2.258	1.833	67.76
	...	XXS	...	0.875	6.875	37.1	21.30	2.258	1.800	72.42
	160	.....	...	0.906	6.813	36.5	21.97	2.258	1.784	74.69

† a. ASA B36.10 steel pipe schedule numbers
  b. ASA B36.10 steel pipe nominal wall thickness designations
  c. ASA B36.19 stainless steel pipe schedule numbers

Table VI Properties of Steel Pipe (Cont'd)

Nominal pipe size, O.D., in.	Schedule number†			Wall thickness, in.	I.D., in.	Inside area, sq. in.	Metal area, sq in.	Sq ft. outside surface, per ft	Sq ft. inside surface, per ft	Weight per ft. lb
	a	b	c							
10 10.750	...	.....	5S	0.134	10.482	86.3	4.52	2.815	2.744	15.15
	...	.....	10S	0.165	10.420	85.3	5.49	2.815	2.728	18.70
	20	.....	...	0.250	10.250	82.5	8.26	2.815	2.683	28.04
	...	.....	...	0.279	10.192	81.6	9.18	2.815	2.668	31.20
	30	.....	...	0.307	10.136	80.7	10.07	2.815	2.654	34.24
	40	Std	40S	0.365	10.020	78.9	11.91	2.815	2.623	40.48
	60	XS	80S	0.500	9.750	74.7	16.10	2.815	2.553	54.74
	80	.....	...	0.593	9.564	71.8	18.92	2.815	2.504	64.33
	100	.....	...	0.718	9.314	68.1	22.63	2.815	2.438	76.93
	120	.....	...	0.843	9.064	64.5	26.24	2.815	2.373	89.20
	140	.....	...	1.000	8.750	60.1	30.6	2.815	2.291	104.13
	160	.....	...	1.125	8.500	56.7	34.0	2.815	2.225	115.65
12 12.750	...	.....	5S	0.165	12.420	121.2	6.52	3.34	3.25	19.56
	...	.....	10S	0.180	12.390	120.6	7.11	3.34	3.24	24.20
	20	.....	...	0.250	12.250	117.9	9.84	3.34	3.21	33.38
	30	.....	...	0.330	12.090	114.8	12.88	3.34	3.17	43.77
	...	Std	40S	0.375	12.000	113.1	14.58	3.34	3.14	49.56
	40	.....	...	0.406	11.938	111.9	15.74	3.34	3.13	53.53
	...	XS	80S	0.500	11.750	108.4	19.24	3.34	3.08	65.42
	60	.....	...	0.562	11.626	106.2	21.52	3.34	3.04	73.16
	80	.....	...	0.687	11.376	101.6	26.04	3.34	2.978	88.51
	100	.....	...	0.843	11.064	96.1	31.5	3.34	2.897	107.20
	120	.....	...	1.000	10.750	90.8	36.9	3.34	2.814	125.49
	140	.....	...	1.125	10.500	86.6	41.1	3.34	2.749	139.68
	160	.....	...	1.312	10.126	80.5	47.1	3.34	2.651	160.27
14 14.000	10	.....	...	0.250	13.500	143.1	10.80	3.67	3.53	36.71
	20	.....	...	0.312	13.376	140.5	13.42	3.67	3.50	45.68
	30	Std	...	0.375	13.250	137.9	16.05	3.67	3.47	54.57
	40	.....	...	0.437	13.126	135.3	18.62	3.67	3.44	63.37
	...	XS	...	0.500	13.000	132.7	21.21	3.67	3.40	72.09
	...	.....	...	0.562	12.876	130.2	23.73	3.67	3.37	80.66
	60	.....	...	0.593	12.814	129.0	24.98	3.67	3.35	84.91
	...	.....	...	0.625	12.750	127.7	26.26	3.67	3.34	89.28
	...	.....	...	0.687	12.626	125.2	28.73	3.67	3.31	97.68
	80	.....	...	0.750	12.500	122.7	31.2	3.67	3.27	106.13
	...	.....	...	0.875	12.250	117.9	36.1	3.67	3.21	122.66
	100	.....	...	0.937	12.126	115.5	38.5	3.67	3.17	130.73
	120	.....	...	1.093	11.814	109.6	44.3	3.67	3.09	150.67
	140	.....	...	1.250	11.500	103.9	50.1	3.67	3.01	170.22
	160	.....	...	1.406	11.188	98.3	55.6	3.67	2.929	189.12
16 16.000	10	.....	...	0.250	15.500	188.7	12.37	4.19	4.06	42.05
	20	.....	...	0.312	15.376	185.7	15.38	4.19	4.03	52.36
	30	Std	...	0.375	15.250	182.6	18.41	4.19	3.99	62.58
	...	.....	...	0.437	15.126	179.7	21.37	4.19	3.96	72.64
	40	XS	...	0.500	15.000	176.7	24.35	4.19	3.93	82.77
	...	.....	...	0.562	14.876	173.8	27.26	4.19	3.89	92.66
	...	.....	...	0.625	14.750	170.9	30.2	4.19	3.86	102.63
	60	.....	...	0.656	14.688	169.4	31.6	4.19	3.85	107.50
	...	.....	...	0.687	14.626	168.0	33.0	4.19	3.83	112.36

† a. ASA B36.10 steel pipe schedule numbers
  b. ASA B36.10 steel pipe nominal wall thickness designations
  c. ASA B36.19 stainless steel pipe schedule numbers

appendix two
ANSWERS TO EVEN-NUMBERED PROBLEMS

CHAPTER ONE

1-2—0.019 ft^3/lb

1-4—(a) 1.88 slug/ft^3
(b) 970 kg/m^3

1-6—195 lb

1-8—1860 lb

CHAPTER TWO

2-2—(a) 32.1 cs
(b) 3.21 × 10^{-5} m^2/s

2-4—(a) 1.6 × 10^{-2} ft^2/s
(b) 14.9 × 10^{-4} m^2/s

2-6—89.8 cp

2-8—0.93 × 10^{-4} m^2/s

CHAPTER THREE

3-2—0.56 gal

3-4—4.7 psi

CHAPTER FOUR

4-2—(a) 13.8 ft^3/lb
(b) 0.07 lb/ft^3

4-4—(a) 12.6 psia
(b) 34.7 psia
(c) 15.7 psia
(d) 14.7 psia

4-6—8.2 psi

4-8—4.99 ft^3

4-10—1666 lb/ft^2

CHAPTER FIVE

5-2—0.98 ft^3
5-6—(a) 1.98 ft^3
    (b) 4.18 ft^3

5-4—66.7 ft^3
5-8—825 m^3

CHAPTER SIX

6-2—19.8 psi
6-6—14.7 psi

6-4—5.5 psi
6-8—1.2 psi

CHAPTER SEVEN

7-2 —(a) 1.42 psi
    (b) 3.28 ft water
    (c) 2.89 in. Hg
7-6 —11.5 ft
7-10—16.65 ft water

7-4—8.8 psia

7-8—12.6 ft water

CHAPTER EIGHT

8-2—(a) 270 ft/s
    (b) 607 ft/s
    (c) 270 ft/s
8-8—56.5 ft/s

8-4 —64.7 ft/s
8-6 —55.9 psi

8-10—0.175 ft^3/s

CHAPTER NINE

9-2—N_R = 2133 but should be less than 2000 for laminar flow
9-4—69.5 ft of water   *9-6*—0.0338
9-8—(a) 1.39 × 10^4
    (b) 2.87 × 10^4

CHAPTER TEN

10-2—0.044 lb/s

10-4—(a) 245 ft/s
     (b) 424 ft/s
     (c) 21.9 lb/s

CHAPTER ELEVEN

11-2—97.4 psi
11-6—198 psi

11-4—30 psi

CHAPTER TWELVE
12-2—5.24 hp

12-6—1.06 hp

12-4—3690 ft lb

CHAPTER THIRTEEN
13-2—(a) 0
(b) No side loadings imposed on the bearings.

13-4—$F_x = 5467$ lb; $F_y = 5568$ lb

CHAPTER FOURTEEN
14-2—10%

14-4—47.2 ft/s

CHAPTER FIFTEEN
15-2—(a) 5.2 ft
(b) 14 ft
(c) 87 ft

15-4—0.7 psi

CHAPTER TWENTY-ONE
21-6—0

CHAPTER TWENTY-TWO
22-6—344 lb in.

CHAPTER TWENTY-THREE
23-4—2.95 ft^2

INDEX

A

Acceleration, 5
  of gravity, 3
Accumulator, 225–226
Acoustic velocity, 101
Actuators
  linear, 165, 175
  rotary, 175
Adiabatic exponent, 42
  values (*see Appendix, Table IV*), 258
Air
  ejection of parts, 162–163
  filtration, 172
  friction loss (*see Friction loss*)
  lubricators for, 172, 230
  pressure regulation, 172
  properties (*see Appendix, Table IV*), 258
Air motors (*see Motors*)
Air-oil system, 169–170
Aluminum, use in tubing, 203
American National Standards Institute, 233
American Society of Mechanical Engineers, 233

B

Bar (*see Appendix, Table II*), 256
Barometer, 26
Bearing loads, in vane pumps, 211
Benzene, properties (*see Appendix, Table I*), 255
Bernoulli's equation, 72–75, 81–82
  for compressible fluids, 95
Boiling, 20
Boosters (*see Intensifiers*)
Bourdon-tube pressure gage, 63–64
Boyle's law, 30–31
Btu, 40–41

C

Capillarity, 19–20
Carbon dioxide, properties (*see Appendix, Table IV*), 258
Carbon tetrachloride, properties (*see Appendix, Table I*), 255
Castor oil, properties (*see Appendix, Table I*), 255

Cavitation, 22, 208
Centrifugal force, in vane pumps, 211
Charles' law, 31
Circuit design, graphic symbols (*see Fluid power, graphic symbols*)
Circuits
  design of, 233
  hydraulic, 243–246
  pneumatic, 249
Coefficient of contraction (*see Orifice*)
Coefficient of discharge (*see Orifice and venturi*)
Coefficient of velocity (*see Orifice*)
Compressor, 29, 34, 39, 113
Conservation of energy (*see Energy*)
Conservation of mass (*see Mass*)
Cooling, of hydraulic fluid, 171
Copper, use in tubing, 203
Corrosion, effect in selecting piping, 201
Cylinders
  cushioned, 181
  functioning and types, 175–179
  horsepower, 115
  mounting, 179–181
  operation in parallel, 251
  piston rod velocity, 166–167
  speed control, 245
  unusual applications, 182–183

D

Darcy equation, 85
Desiccator, 252

E

Efficiency, 117–118
  volumetric (*see Motors*)
Electric motors, compared to hydraulic, 220
Elevation head, 74–76
Energy
  additions and losses, 75
  conservation, 69, 72–75
  units, 40–41
Equation of continuity
  compressible fluids, 94
  incompressible fluids, 70
Equivalent length of pipe (*see Friction loss*)
Ethyl alcohol, properties (*see Appendix, Table I*), 255
Exhaust, of hydraulic and pneumatic systems, 166

F

Fatigue, in copper tubing, 203
Felt, for filters, 228
Filters, 228–230
  materials for, 228
  rating of, 228
Filtration, multi-stage, 228
Fittings, for tubing and hose, 205
Flow
  steady state, 69
  through orifices (*see Orifice*)
  through venturi (*see Venturi*)
Flow types
  laminar, 82–83
  turbulent, 83
Fluid dynamics
  compressible fluids, 93–100
  incompressible fluids, 69–78
Fluid filtration, 171
Fluid mechanics, definition, 1
Fluid power, 1, 153
Fluid power applications, 153–163
  agricultural implement, 154
  aircraft, 155
  manufacturing, 157, 159–160
  marine, 158
  missiles, 156
  printing, 160
  trucks, 155, 157
Fluid power, graphic symbols, 233–240
Fluids
  compressible, 2 (*see also Gases*)
  definition, 2
  ideal, 2
  incompressible, 2 (*see also Liquids*)
  properties, 2–4
Fluid statics, 47–53
Foot pound, 40, 111
Force
  caused by a jet on a flat surface, 124–127
  caused by a jet on a moving surface, 127–128
  mass-acceleration relationship, 4
  output of a cylinder, 13, 115
  on pipe bend, 129–133
  shear, 7–8
Friction factor, 87–88
Friction loss, 40, 75, 85–87
  air, 103–109
  entrances, 145, 149–150
  fittings and valves, 145–151
  L/d ratios, table, 147
Fritted glass, for filters, 228

INDEX

G

Gas constant, values (*see Appendix, Table IV*), 258
Gases, 25
    kinetic theory, 30
    perfect gas law, 32
    pressure-height relationship, 53–54
    volume at standard conditions, 94
    universal gas constant, 32
Gasoline, properties (*see Appendix, Table I*), 255
Glycerine, properties (*see Appendix, Table I*), 255
Graphic symbols (*see Fluid power, graphic symbols*)

H

Harris formula, 104–106
Heat, 40–41
    transfer, 42
Horsepower (*see Power*)
Hose, 204
Hydraulics, 25, 161, 165–167
Hydraulic vs. pneumatic system, 166–167
Hydrogen, properties (*see Appendix, Table IV*), 258
Hydrostatic bearings, 161–163
Hydrostatic transmissions, 217

I

Impulse, impulse-momentum principle, 121–133
Impulse turbine, 123
Intensifiers, 205–206

J

Jet
    action on a turbine blade, 123, 127–129
    impingement on stationary surfaces, 124–127
JIC specifications, 203
Joule, 40–41

K

Kerosene, properties (*see Appendix, Table I*), 255
K-factor, for pipe resistances, 149–150
Kilogram, 4
Kilopascal, 58
Kinetic energy, 72–73

L

Laminar flow, in aircraft design, 153
Linseed oil, properties (*see Appendix, Table I*), 255
Liquids
    compressibility, 21–22
    properties, 17–22
Lubrication
    in hydraulic systems, 170
    in pneumatic systems, 172

M

Mach number, 100–101
Manometer, 26, 59–63
    differential, 60
Mass
    conservation of, 69–71
    definition, 2
Media, 228
Meniscus, 19
Mercury, properties (*see Appendix, Table I*), 255
Methane, properties (*see Appendix, Table IV*), 258
Methyl alcohol, properties (*see Appendix, Table I*), 255
Metric system, equivalent units (*see SI system*)
Micrometer, 228
Micron, 228
Mineral oil, properties (*see Appendix, Table I*), 255
Mole, 35
Molecular weight, 35
    of gases (*see Appendix, Table IV*), 258
Momentum (*see Impulse-momentum principle*)
Moody's chart, 86–90
Motor oil, properties (*see Appendix, Table I*), 255
Motors
    air, 216–217, 220–221
    axial piston, 219–220
    comparison of hydraulic and electric, 220
    gear, 219–220
    hydraulic, 218–220
    torque and power, 215–217
    vane, 219–220
    variable displacement, 222
    volumetric efficiency, 216

N

NBS flow factor, 106–108
Newton, 9
Nitrogen, properties (*see Appendix, Table IV*), 258
NPTF threads, 203
NPT threads, 203
Nylon, use in tubing, 203

O

Oil level gage for reservoirs, 231
Orifice, 135–138
    coefficient of contraction, 136
    coefficient of discharge, 137
    coefficient of velocity, 136–137
    use as a flow quantity measuring device, 135
    vena contracta, 136
Oxidation, of oil, 203
Oxygen, properties (*see Appendix, Table IV*), 258

P

Paper, for filters, 228
Pascal, 9–10, 28
Pascal's law, 7, 12–15
Phosphate ester oil, properties (*see Appendix, Table I*), 255
Pipe fittings, friction loss (*see Friction loss*)
Pipe roughness, 87–88
Piping
    allowable stresses, 203
    factors in selection of, 201–203
    steel, 202
    table of properties (*see Appendix, Table VI*), 259
Pitot tube, 139–140
Plastic, use in tubing, 203
Pneumatic system vs. hydraulic (*see Hydraulic vs. pneumatic system*)
Pneumatics, 25, 161, 167–168
Poise, 8–10
Polyethylene, use in tubing, 203
Polyvinyl chloride, use in tubing, 203
Potential energy, 72
Potential head (*see Elevation head*)
Pound, unit of weight, 4
Powder metals for filters, 228
Power
    electrical, 111
    output of a cylinder, 115

Power, units of, 111
Power steering, automobile, 154
Pressure,
    absolute, 27
    atmospheric, 25–28
    force-area relationship, 13
    gage, 27
    head, 49–50, 74–75
    height relationship, 47–49
    measurement, 25–28, 57–58
        units of measurement (*table*), 58
Pressure gages, Bourdon-tube, 63–64
Pumps
    cavitation, 208
    characteristics, 212–213
    gear, 209
    horsepower, 112–114
    hydrodynamic, 207–208
    piston, 211–212
    positive displacement, 207
    size range, 213
    types, 207–208
    vane, 210

R

Reservoirs, 230–232
    design considerations, 231–232
    operation in cold temperatures, 232
Reynolds number, 83–84

S

Saybolt universal viscosimeter, 9–10
Scfm, 94
Seawater, properties (*see Appendix, Table I*), 255
Shop air system, water and sludge in, 172
Silicone oil, properties (*see Appendix, Table I*), 255
SI system, equivalent units with English system (*see Appendix, Table II*), 256
Slug, 4
Specific gravity
    definition, 3
    values for common liquids (*see Appendix, Table I*), 255
Specific volume, definition, 3
Specific weight
    definition, 3
    values for common liquids (*see Appendix, Table I*), 255
    values for gases (*see Appendix, Table IV*), 258

INDEX

SSU seconds, 9-10
Stoke, 9-10
Strainers, 227
Stress, shear, 8
Surface tension, 17-19
    values for various liquids, 18
Systems of units
    CGS metric, 4
    English, 4
    SI, 4

T

Temperature
    absolute, 28
    Celsius scale, 3, 28
    Fahrenheit scale, 28
    Kelvin scale, 28
    limits for tubing, 203
    Rankin scale, 28
Tetrafluoroethylene, use in tubing, 203
Thermodynamic processes
    adiabatic, 42-43, 96
    isothermal, 41-42
    polytropic, 43
Threads, 203
Torque, 215-217
Torricelli's theorem, 78
Tubing
    allowable pressures, 203-204
    factors in selection, 201, 203-205
    friction loss, 106-108
    material, 203
    wall thickness, 204
Turbine (*see Impulse turbine*)

U

Universal gas constant (*see Gases*)

V

Vacuum, definition, 27
Valves
    application of directional control, 190-191, 193
    check, 192
    definition, 185
    directional control, 185-190

Valves (*continued*)
    flow control, 196
    gate, 186
    globe, 186
    needle, 187
    pilot operation, 196
    plug, 186
    pressure control, 192-196
    relief, 193-195
    solenoid operated, 190, 197
Vapor lock, 21
Vapor pressure, 20-21
    values for common liquids (*see Appendix, Table I*), 255
Vectors, 121-122
Velocity
    flow, 70-71
    head, 74-76, 95-96
    in Reynolds number, 83
    in viscosity determination, 8
    of free jet, 78
    of sound, 101
Vena contracta (*see Orifice*)
Venturi, 138-139
    coefficient, 139
Viscosity, 2
    absolute, 8
    conversion of units, 10-11
    definition, 7
    index, 12
    kinematic, 9
    table of values (*see Appendix, Table III*), 257
Volumetric efficiency (*see Motors*)

W

Water, properties (*see Appendix, Table I*), 255
Water-glycol solution, properties (*see Appendix, Table I*), 255
Water hammer, 225-226
Water-oil emulsion, properties (*see Appendix, Table I*), 255
Wire mesh and cloth, for filters, 228

Y

Yarn, for filters, 228